Autocad 2000i

An Introductory Course

Autocad 2000i

An Introductory Course

Ian Mawdsley

Newnes

OXFORD   AUCKLAND   BOSTON
JOHANNESBURG   MELBOURNE   NEW DELHI

Newnes

An imprint of Butterworth-Heinemann
Linacre House, Jordan Hill, Oxford OX2 8DP
225 Wildwood Avenue, Woburn, MA 01801–2041
A division of Reed Educational and Professional Publishing Ltd

℞ A member of the Reed Elsevier plc group

First published 2001

British Library Cataloguing in Publication Data

A catalogue record of this book is available from the British Library

ISBN 07506 47221

⚱ Typeset and produced by Sylvester Publications (Loughborough and Sunderland)

Printed and bound in Great Britain

FOR EVERY TITLE THAT WE PUBLISH, BUTTERWORTH-HEINEMANN
WILL PAY FOR BTCV TO PLANT AND CARE FOR A TREE.

Contents

Preface   *vii*

1   Getting started   1

2   Basic drawing tools   19

3   Drawing aids   45

4   Basic editing tools   61

5   Manipulating the drawing view   93

6   More draw commands   101

7   More modify commands   117

8   Layers, linetypes and colours   123

9   Text   135

10 Hatching   151

11 Dimensioning   161

12 Templates   187

13 Modifying object properties   195

14 Grips   201

15 Blocks   207

16 Attributes   219

17 Model space and Paper space   227

18 AutoCAD DesignCenter   241

19 Communicating between drawings   251

20 The User Co-ordinate System   259

21 Modelling using 3D solids   263

Index   *291*

Preface

The aim of this book is to introduce the user to the 2D drawing, editing and management features of the AutoCAD 2000i and AutoCAD LT2000i systems. Also to appreciate the basics of the 3D modelling features.

The book has been structured to develop the broad level of understanding required to appreciate the essential concepts and principles associated with a professional CAD system and is suitable for those with little or no understanding of a CAD system.

The information presented within this book is intended to be used as an interactive teaching aid. Learning will take place through the completion of worked examples, exercises and assignments.

The assignments and exercises have been structured to develop the skills necessary to undertake the City & Guilds 4351 Computer Aided Draughting and Design series and will also benefit students undertaking the EdExcel Advanced Award in AutoCAD.

It is not the purpose of this book to explore every facet of the commands contained within the entire AutoCAD 2000i system, rather it is to make the reader aware of the variety of commands that are at your disposal. Successful completion and understanding of the work contained within the book will prepare you for progression to more advanced work using the AutoCAD system.

Chapter 1

Getting started

The object of this section is to introduce the user to the graphics screen, toolbar and toolbox, also the features that are undertaken at the start of a drawing. These include how to start a new drawing, setting up the drawing limits, units, grid and snap values. A summary of the drawing commands is included to give an overview of their features prior to their use.

Objectives

At the end of this chapter you will be able to:

▷ Start a new drawing.
▷ Recognise the various features of the user interface.
▷ Understand the principle of the command line window.
▷ Select and position toolbars.

New commands

▷ Polyline   PL   ↵

▷ Help   ?

AutoCAD Today

The AutoCAD Today dialogue box will automatically activate when you open AutoCAD or start a new drawing.

The AutoCAD Today dialogue box is split into three areas.

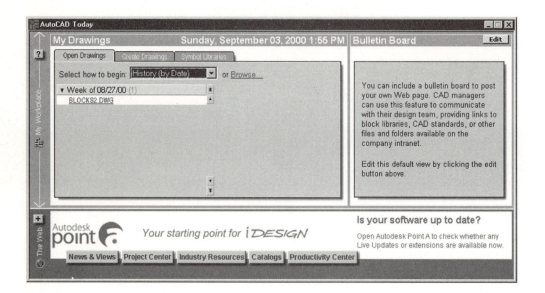

My Drawings

- will connect you to drawings on your computer or network,
- allow you to start new drawings from scratch or by using a template, or
- access symbol libraries from existing files located on your computer or network.

Bulletin Board

The Bulletin Board allows a CAD manager to display information that may be required by the design team.

Autodesk point A

Provides industry-specific news, resources, links, and other services through the internet. Autodesk Point A can only be accessed if you are currently connected to the web.

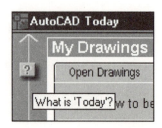

Select the What is 'Today'? icon to display the Today information window.

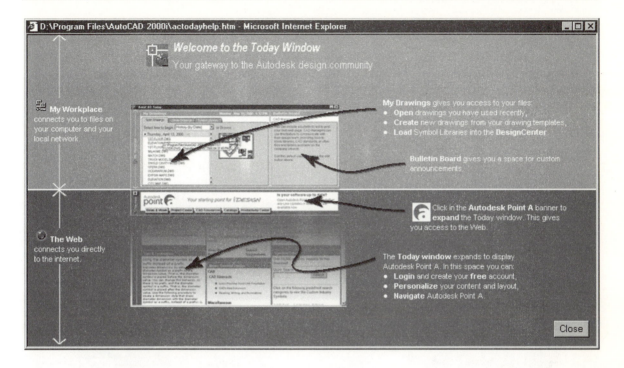

My Drawings section

Use this section to access or start a new drawing.

Open Drawings tab

The Open Drawings tab allows you to select a drawing using one of four options.

- Most recently used
- History (by date)
- History (by filename)
- History (by location)

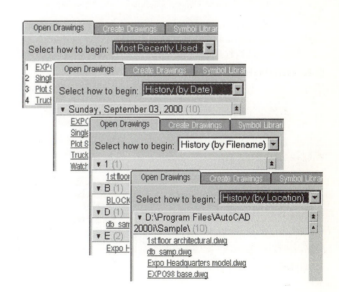

Or you can use the Browse option to locate the required file.

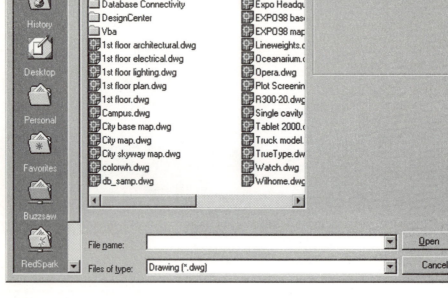

Create Drawings tab

The Create Drawings tab allows you to create a new drawing using one of three options.

Template

You can start a new drawing quickly with a template drawing. A template drawing is an AutoCAD drawing file that has the extension *.dwt* and is saved automatically in the template directory.

Template drawings can store AutoCAD settings that you would otherwise have to set individually. Template drawings are also useful for maintaining drawing standards and conventions. In a template drawing file you can predefine many settings including:

- Drawing units
- Sheet size and layout
- Text and dimensioning styles
- Colours
- Types of lines
- Layer names
- View names
- UCS names
- Title blocks

Start from Scratch

If you want to begin drawing quickly using default English or Metric settings choose *Start from Scratch*.

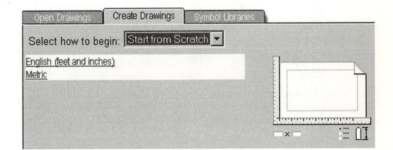

Wizards

If you want help to set up a drawing, choose *Wizards*. You can choose from two wizards: *Quick Setup* or *Advanced Setup*.

Quick Setup

Choose the Quick Setup wizard to select your units type, (i.e. Decimal) and drawing area.

Advanced Setup

Use Advanced Setup to set up units, angle variables, drawing area, title block and drawing layout.

Symbol Libraries tab

Using the Symbol Libraries tab will start a new drawing and also provide a library of symbols that can be inserted into the drawing from the DesignCenter (see Chapter 18).

Creating drawings using pre-prepared symbol libraries greatly reduces drafting time and maintains standardisation across a range of similar drawing files.

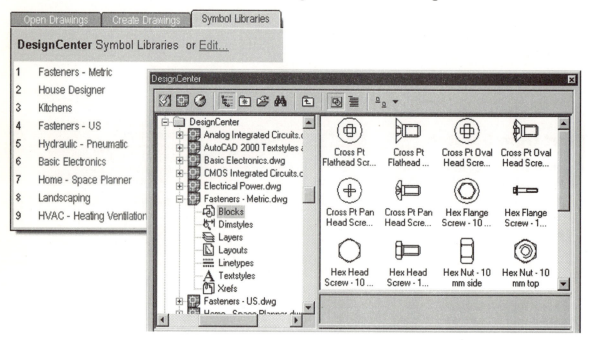

The Edit option activates the Symbol Libraries dialogue box. Here, you can locate libraries using the Browse button and re-order libraries in order of preference.

Bulletin Board section

The Bulletin Board allows a CAD manager to communicate information that may be required by the design team such as memos, design alterations, symbol libraries etc.

You can create a bulletin board message using a word processor such as Word or Notepad.

Example

Creating a bulletin board message

1.   Open the Notepad word text editor.
2.   Enter the following text:
Memo – Design Team
Urgent meeting 1.15pm Rm 101.
3.   Save your file as *Memo.txt* in the Temp folder.
You can specify a path and file name to the bulletin board file using the Bulletin Board Edit button (the default bulletin board file is cadmgr.htm).
4.   Press the Edit button on the bulletin board.
5.   Press the Browse button and navigate to the Temp folder and select the file Memo.txt. Now press the Save Path button. Your message will appear in the Bulletin Board.

Bulletin Board   Edit

You can include a bulletin board to post your own Web page. CAD managers can use this feature to communicate with their design team, providing links to block libraries, CAD standards, or other files and folders available on the company intranet.

Edit this default view by clicking the edit button above.

D:\Program Files\AutoCAD 2000i\editcadmgr.htm - Microsoft Internet Exp...

CAD managers can communicate directly with their local user base throughout the day with messages, reminders, calendars, and links to things like standards and content libraries stored locally.

Enter Path to the Local Server:

D:\TEMP\Memo.txt   Browse...

Save Path   Cancel

Bulletin Board   Edit

Memo - Design Team

Urgent meeting 1.15pm Rm 101

Autodesk Point A

Autodesk Point A is a design resource for the AutoCAD community. It provides industry-specific news, resources, links and other services by connecting you to the web. If your Internet is switched on it helps you to integrate the Internet into your professional work life.

Click the Autodesk Point A icon to access the website.

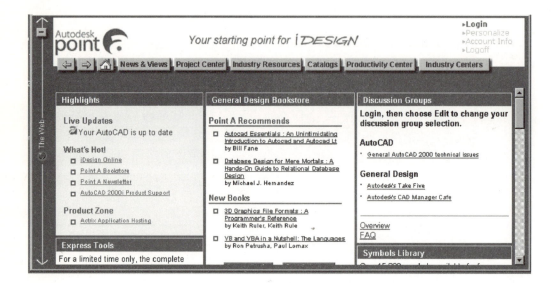

Press the *Create Drawings* tab. Select the *Start from Scratch – Metric* option. The *Today* window will disappear.

Whilst the *Today* window contains a number of useful features, you are provided with an alternative startup dialogue box that is used solely to open or start new drawings.

1. Right-click anywhere within the graphics screen and select Options from the bottom of the short cut menu.

The Options dialogue box contains all the settings and user preferences within the AutoCAD system.

2. Select the System tab.

Under the General Options area you will see that the Startup: option is set to:

'*Show TODAY startup dialog*'.

Use the down arrow and select the Startup: option:

'*Show traditional startup dialog*'.

3. Press the Apply button followed by the OK button.

In order to demonstrate that the traditional startup dialogue box is now set current, you will now exit the AutoCAD system.

4. Select Exit from the *File...* pull down menu to close the system down.

5. Activate the AutoCAD 2000i.

When you start the AutoCAD system you will be presented with the *Startup* dialogue box. You are provided with the following options:

Open a Drawing

Use this option to open an existing drawing.

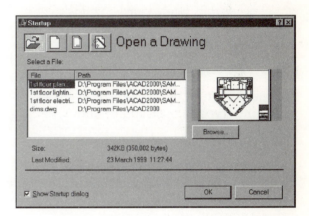

Start from Scratch

If you want to begin drawing quickly using default English or metric settings choose *Start from Scratch*.

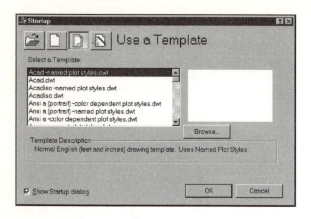

Use a Template

You can create new drawings quickly using the template option.

Use a Wizard

If you want to set up a drawing using a dialogue box, choose Use a Wizard. You can choose from two wizards: *Quick Setup* or *Advanced Setup*.

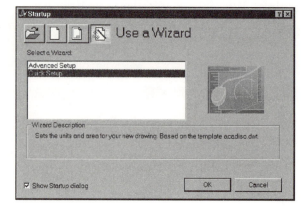

Quick Setup

Choose the Quick Setup wizard to select your units type, (i.e. Decimal) and drawing area.

Advanced Setup

Use Advanced Setup to set up units, angle variables, drawing area, title block and drawing layout.

> **Note** – for the purpose of this book we will continue to use this startup dialogue box. From the *Start Up* dialogue box pick the *Start from Scratch* button and select the *Metric* option, then *OK.*

Active Assistance

When you open the AutoCAD 2000i system, the Active Assistance dialogue box will appear.

Active Assistance provides information relating to the command currently in use and updates dynamically as new commands are activated.

Active Assistance options

Right-click anywhere within the Active Assistance dialogue box to activate the short cut menu.

Home: Displays the original Using the Active Assistance window.

Back: Displays the previous command.

Forward: Scrolls commands forward after using the Back: option.

Print: Prints the current Active Assistance window contents.

Settings...

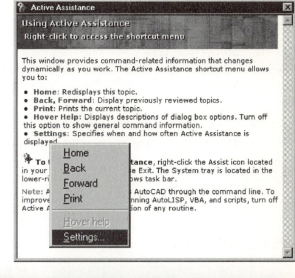

Show on start

This option ensures that the Active Assistance will open when AutoCAD starts.

Hover Help

When selected, Active Assistance displays information about specific features within an open dialogue box when the cursor passes over them.

Activation

This option determines when the Active Assistance window will open.

All commands

When any command is activated.

New and enhanced commands

When any new or enhanced commands are activated.

Dialogues only

When a dialogue box is displayed.

On demand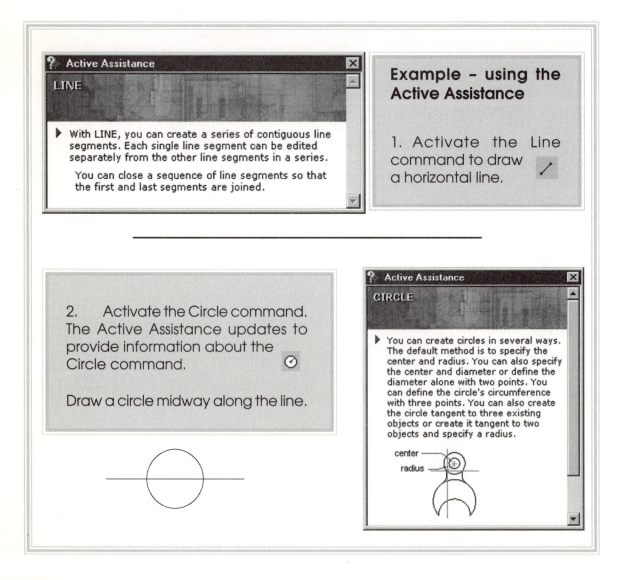

This option stops the automatic display of the Active Assistance window. To open the window, double-click the Active Assistance icon.

> **Note – since the Active Assistance does take up a portion of the graphics screeen, it may get in the way of your drawing. You may consider setting the Activation – on demand option.**

Active Assistance

LINE

▶ With LINE, you can create a series of contiguous line segments. Each single line segment can be edited separately from the other line segments in a series.

You can close a sequence of line segments so that the first and last segments are joined.

Example – using the Active Assistance

1. Activate the Line command to draw a horizontal line.

2. Activate the Circle command. The Active Assistance updates to provide information about the Circle command.

Draw a circle midway along the line.

Active Assistance

CIRCLE

▶ You can create circles in several ways. The default method is to specify the center and radius. You can also specify the center and diameter or define the diameter alone with two points. You can define the circle's circumference with three points. You can also create the circle tangent to three existing objects or create it tangent to two objects and specify a radius.

center
radius

3. Activate the Linear dimension command in the Dimension pull-down menu.
Active Assistance now provides linear dimension information

Right-click the Active Assistance box and select Settings...
Activate *Show on Start* and set Activation – *On Demand* then OK.

The Graphics Window

A graphics screen similar to the illustration below will appear.

The graphics screen contains a menu bar, toolbars, drawing area, graphics cursor, command line and status bar.

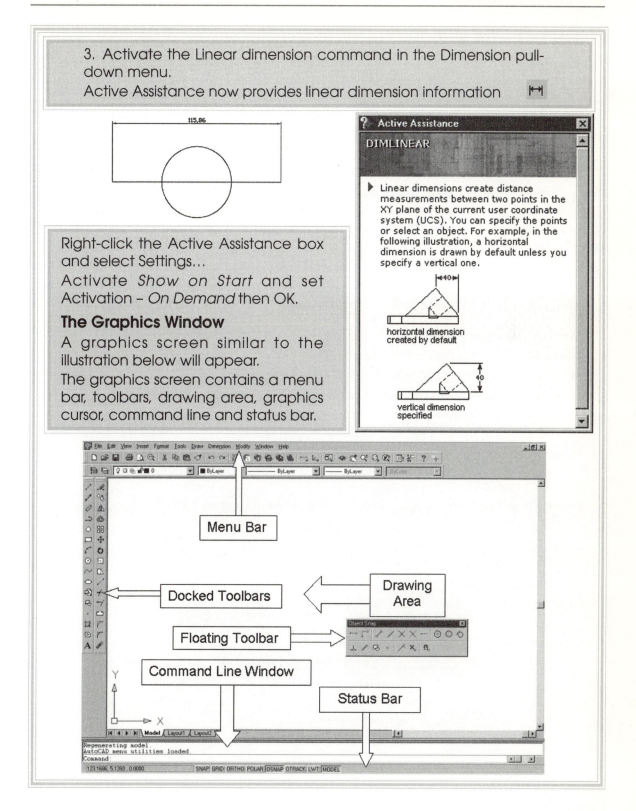

Active Assistance

DIMLINEAR

▶ Linear dimensions create distance measurements between two points in the XY plane of the current user coordinate system (UCS). You can specify the points or select an object. For example, in the following illustration, a horizontal dimension is drawn by default unless you specify a vertical one.

horizontal dimension created by default

vertical dimension specified

Menu Bar

Docked Toolbars

Drawing Area

Floating Toolbar

Command Line Window

Status Bar

Menu bar and pull-down menus

The menu bar is situated at the top of the screen. The menu bar consists of a series of menu titles. These pull-down menus provide quick reference to commonly used commands and features.

File	Edit	View	Insert	Format	Tools	Draw	Dimension	Modify	Express	Window	Help

Command line window

You can enter commands on the command line by typing the name of the command at the Command prompt using the keyboard. A command will either display a set of options or activate a dialogue box. A command is executed by pressing the spacebar, or enter/return button.

Example – polyline command

At the Command prompt enter *PLINE* (for Polyline) then press the RETURN button (⏎).
Specify start point: *enter* 20,20.
A rubber-band line will appear on your screen connected to the cursor and a set of options will appear at the command line:
Current line-width is 0.0000.
Specify next point or (Arc/Close/Halfwidth/Length/Undo/Width):
You can click your cursor on the drawing screen to indicate the endpoint of the line or you can enter a new co-ordinate position at the command line.
Enter the new co-ordinates *50,50*⏎.
Press the RETURN button a second time to end the command.
To activate any of the current command options, enter the uppercase letter of the required option.

```
AutoCAD Text Window - Drawing1
Edit
Command:
PLINE
Specify start point:

Current line-width is 0.0000
Specify next point or [Arc/Close/Halfwidth/Length/Undo/Width]:
Command: |
```

Aliases

Most of the commands that can be entered at the command line will also have an alias. An alias is an abbreviated command or short cut entry. For example, the alias for the Line command is L. Other aliases include, PL for Polyline command, C for Circle command, M for Move and E for Erase.

Toolbars

AutoCAD has a comprehensive set of toolbars. Choosing a tool on a toolbar will activate the command associated with that tool.

The following toolbars are generally visible when you start a new drawing, although you can turn them off if you wish.

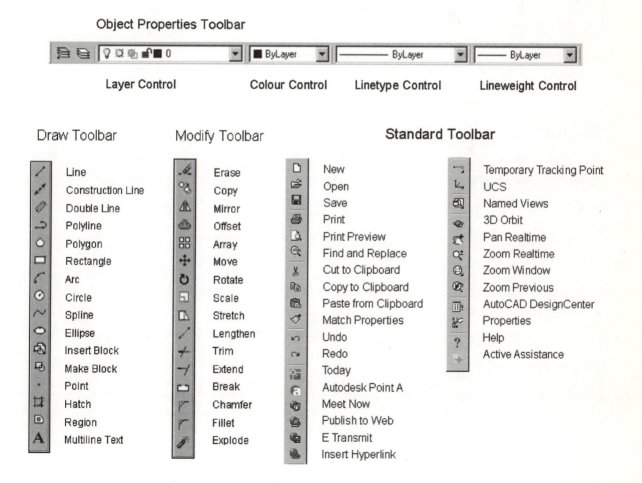

Object Properties Toolbar

Layer Control   Colour Control   Linetype Control   Lineweight Control

Draw Toolbar   Modify Toolbar   Standard Toolbar

Draw Toolbar	Modify Toolbar	Standard Toolbar	
Line	Erase	New	Temporary Tracking Point
Construction Line	Copy	Open	UCS
Double Line	Mirror	Save	Named Views
Polyline	Offset	Print	3D Orbit
Polygon	Array	Print Preview	Pan Realtime
Rectangle	Move	Find and Replace	Zoom Realtime
Arc	Rotate	Cut to Clipboard	Zoom Window
Circle	Scale	Copy to Clipboard	Zoom Previous
Spline	Stretch	Paste from Clipboard	AutoCAD DesignCenter
Ellipse	Lengthen	Match Properties	Properties
Insert Block	Trim	Undo	Help
Make Block	Extend	Redo	Active Assistance
Point	Break	Today	
Hatch	Chamfer	Autodesk Point A	
Region	Fillet	Meet Now	
Multiline Text	Explode	Publish to Web	
		E Transmit	
		Insert Hyperlink	

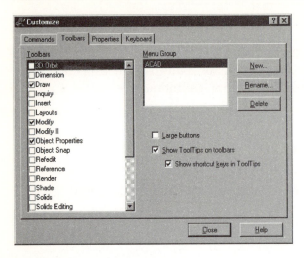

You can have as many toolbars visible on the screen as you wish but they do take up space and reduce the available drawing area.

The *Toolbars* dialogue box that displays or hides toolbars can be activated by selecting *Toolbars...* in the *View* pull-down menu.

You activate a toolbar by selecting the box to the left of its name.

Floating toolbars

Floating toolbars can be either docked at the edges of the screen or placed in any position on the screen. When the cursor is placed over a tool the name of the tool is displayed by a ToolTip below the cursor.

Status bar

The status bar at the bottom of the graphics screen displays the current cursor co-ordinates and is used to activate:

GRID, SNAP, ORTHO, POLAR, OSNAP, OTRACK, LWT and MODEL/PAPER modes.

| 150.1546, 358.5709, 0.0000 | SNAP | GRID | ORTHO | POLAR | OSNAP | OTRACK | LWT | MODEL |

A depressed button indicates that the mode is active. A single click on the mode button will activate the mode.

Help command

The AutoCAD Help system contains complete information for using AutoCAD.

To activate the AutoCAD Help system choose *AutoCAD Help Topics* from the *Help* menu or press the Help button on the Standard Toolbar.

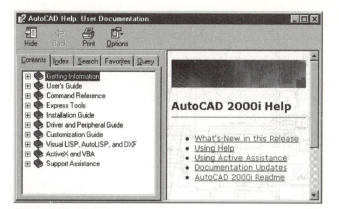

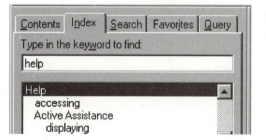

Select the *Index* tab and type in the name of the command or feature you require help with.

Natural language query

The Query tab allows you to use complete sentences or phrases to search and retrieve information.

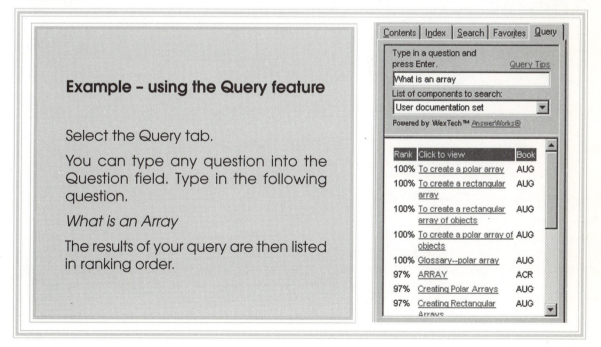

Example – using the Query feature

Select the Query tab.

You can type any question into the Question field. Type in the following question.

What is an Array

The results of your query are then listed in ranking order.

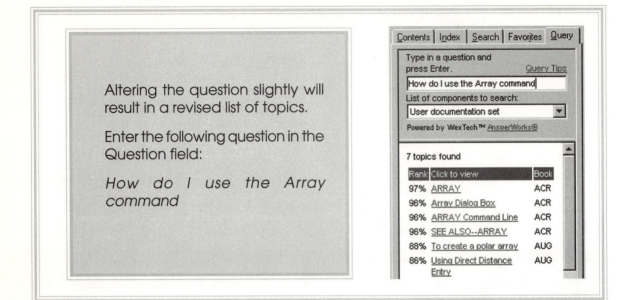

Altering the question slightly will result in a revised list of topics.

Enter the following question in the Question field:

How do I use the Array command

Chapter 2

Basic drawing tools

This chapter covers the essentials of data co-ordinate entry. It is normally taken for granted that when you input an X and Y co-ordinate, you are using the lower left corner as the 0,0 datum, and for the majority of co-ordinate entries this may be suitable. There are circumstances however, especially in complex drawings, where the user may need to take the current co-ordinate position as the 0,0 datum. This chapter introduces the user to the Absolute co-ordinate system (default), the Relative co-ordinate system and the Polar co-ordinate system of point entry. Various exercises using the basic drawing commands, Polyline, Polygon and Circle will illustrate how effective use of each of these co-ordinate entry systems can aid the drawing process. The chapter concludes with a practical assignment designed to test your understanding of the features covered.

Objectives

At the end of this chapter you will be able to:

▷ Understand the principle of Absolute co-ordinate entry.
▷ Understand the principle of Relative co-ordinate entry.
▷ Understand the principle of Polar co-ordinate entry.
▷ Use a variety of basic drawing commands.
▷ Place objects at precise co-ordinate locations.

New commands

▷	Save	SA		▷	Polygon	PG	
▷	Polyline	PL		▷	Line	L	
▷	Rectangle	REC		▷	Circle	C	

Line command

The most basic of all the drawing objects. The initial *from point: (1)* can be picked using the screen cursor or by entering an *x,y* co-ordinate, followed by a *to point: (2)* location.

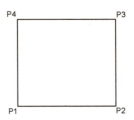

Example

From point:	*enter*	25,25	⏎	(P1)
To point:	*enter*	50,25	⏎	(P2)
To point:	*enter*	50,50	⏎	(P3)
To point:	*enter*	25,50	⏎	(P4)
To point:	*enter*	C	⏎	(to close)

Draw toolbar	
Draw menu	**Line**
Command line entry	**Line**
Alias	**L**

Note – if you make a mistake entering a new line location, press the U key. This will undo your last input without exiting the command.

Undo/redo command

Typing *Undo* at the command prompt will allow you to undo completely your previous command. The *Redo* command will enable you to redo an undo command.

Note – whilst you can undo back as many commands as you like, you can redo forward only once.

Direct distance entry

This is the simplest way to place a series of lines without entering co-ordinate values. You determine your next line point by moving the cursor in the direction you want to go and then enter the required distance from your current position.

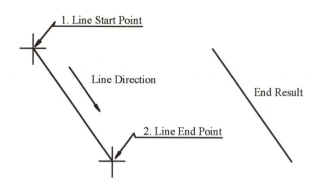

1. Line Start Point

Line Direction

End Result

2. Line End Point

To draw a line using direct distance entry:

1.   Start the Line command and specify the first point (anywhere on the screen).

2.   Move the screen cursor in the direction you want to draw the line.

3.   Specify a distance on the command line and press Return.

The line is drawn at the length entered at the command line and angle specified by the cursor direction.

When *Ortho* mode is on, this is the quickest way to draw horizontal and vertical lines.

Activate the *Ortho* command pressing F8 or by pressing *Ortho* on the status bar.

Repeat instructions 1 to 3.

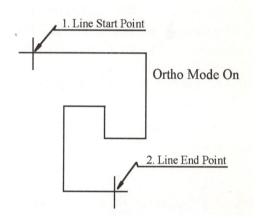

1. Line Start Point

Ortho Mode On

2. Line End Point

1.   Start the *line* command and specify the first point (anywhere on the screen).

2.   Move the screen cursor in the direction you want to draw the line.

3.   Specify a distance on the command line and press Return.

Now you have a go

Using only Direct Distance entry with Ortho mode on, produce the following drawing (without the dimensions!). Start at the lower left corner and go clockwise.

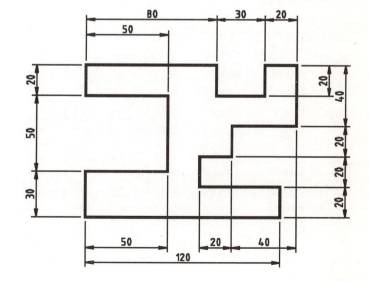

Co-ordinates entry

The AutoCAD system provides you with a variety of methods to locate a point on the drawing screen in order to carry out a particular action, for instance, to locate the endpoint of a line or to move an object to a new location. The most common methods of co-ordinate entry are covered below.

Absolute co-ordinates

Imagine your screen is like a piece of graph paper and all your co-ordinates are taken from one datum base point, the lower left corner, X0, Y0. A co-ordinate position specified from the keyboard will be taken relative to this *X0, Y0* position.

For instance, point *A* shown is 20 units in the X axis and 20 units in the Y axis from datum 0,0, whilst point *B* shown is 70 units along the X axis and 50 units along the Y axis

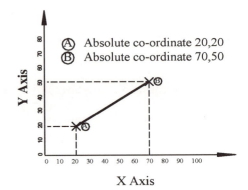

Ⓐ Absolute co-ordinate 20,20
Ⓑ Absolute co-ordinate 70,50

Example

```
Command: Line
From point:  enter    50,50      ↵   (P1)
To point:    enter    50,100     ↵   (P2)
To point:    enter    80,130     ↵   (P3)
To point:    enter    130,130    ↵   (P4)
To point:    enter    160,100    ↵   (P5)
To point:    enter    180,100    ↵   (P6)
To point:    enter    180,70     ↵   (P7)
To point:    enter    90,70      ↵   (P8)
To point:    enter    90,50      ↵   (P9)
To point:    enter    C          ↵   (close)
```

Relative co-ordinates

Relative co-ordinate entry allows you to specify the X,Y co-ordinate of the end point of your next line relative to your current position. In order to inform AutoCAD that you wish to use a relative co-ordinate entry, you must type the @ symbol prior to your next co-ordinate position.

The format is: @ X,Y

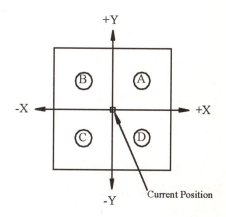

> Where: anywhere in quadrant A = @ X,Y
> (Positive X and Y values),
> Quadrant B = @ –X,Y
> (Negative X value, positive Y value).
> Quadrant C = @ –X,–Y
> (Negative X and Y values
> Quadrant D = @ X,–Y
> (Positive X value, negative Y value).

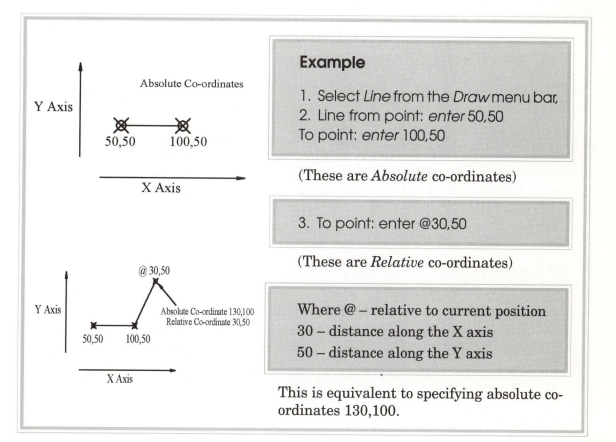

Example

1. Select *Line* from the *Draw* menu bar,
2. Line from point: *enter* 50,50
To point: *enter* 100,50

(These are *Absolute* co-ordinates)

3. To point: enter @30,50

(These are *Relative* co-ordinates)

Where @ – relative to current position
30 – distance along the X axis
50 – distance along the Y axis

This is equivalent to specifying absolute co-ordinates 130,100.

Example

```
Command: Line
From point:    enter    50,50      ⤶    (P1)
To point:      enter    @0,50      ⤶    (P2)
To point:      enter    @30,30     ⤶    (P3)
To point:      enter    @50,0      ⤶    (P4)
To point:      enter    @30,-30    ⤶    (P5)
To point:      enter    @20,0      ⤶    (P6)
To point:      enter    @0,-30     ⤶    (P7)
To point:      enter    @-90,0     ⤶    (P8)
To point:      enter    @0,-20     ⤶    (P9)
To point:      enter    C          ⤶    (close)
```

Polar co-ordinates

You can use Polar co-ordinates to specify your next line point as a Distance and Angular Direction from your current position. Again, in order to inform AutoCAD that you wish to use a relative co-ordinate entry, you must type the @ symbol prior to your next co-ordinate position.

The format is: @ *distance* < *angle*

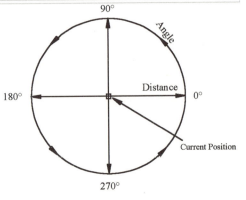

Note – angle 0 (zero) starts at 3 o'clock and continues around in an anti-clockwise direction.

For instance: @30<280

> where: @ – Relative to current position
> 30 – Specifies Polar Distance
> 280 – Angle in Degrees

This would result in a line 30 units long at an angle of 280 degrees from P3 to P4 shown.

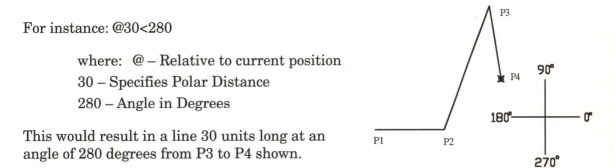

Example

```
Command: Line
From point:     enter     50,50          ⏎     (P1)
To point:       enter     @50<90         ⏎     (P2)
To point:       enter     @30,30         ⏎     (P3)
To point:       enter     @50<0          ⏎     (P4)
To point:       enter     @30,–30        ⏎     (P5)
To point:       enter     @20<0          ⏎     (P6)
To point:       enter     @30<270        ⏎     (P7)
To point:       enter     @90<180        ⏎     (P8)
To point:       enter     @20<270        ⏎     (P9)
To point:       enter     C              ⏎     (close)
```

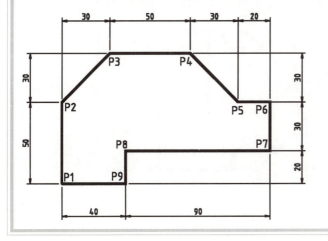

Note – relative co-ordinate entry is used for points (P3) and (P5) because even though we know the angle, we do not know the exact distance.

Example

You will now use a combination of both the Relative and Polar co-ordinate systems to produce the following drawing.

Select *New* from the *File...* pull-down menu. Choose the *Use a Wizard* option and select *Quick Setup*.

Ensure your units are set to *Decimal* and enter an Area width 300 by length 200 then press *Done*. Select *Line* from the *Draw* pull-down menu.

Command: Line				
From point:	*Start at P1 shown*			
To point:	*enter*	@30<0	↵	(P2)
To point:	*enter*	@40,–40	↵	(P3)
To point:	*enter*	@20<0	↵	(P4)
To point:	*enter*	@40,40	↵	(P5)
To point:	*enter*	@30<0	↵	(P6)
To point:	*enter*	@80<270	↵	(P7)
To point:	*enter*	@40<0	↵	(P8)
To point:	*enter*	@30<270	↵	(P9)
To point	*enter*	@80<180	↵	(P10)
To point:	*enter*	@-40,40	↵	(P11)
To point:	*enter*	@-40,–40	↵	(P12)
To point:	*enter*	@80<180	↵	(P13)
To point:	*enter*	@30<90	↵	(P14)
To point:	*enter*	@40<0	↵	(P15)
To point:	*enter*	C	↵	(to close)

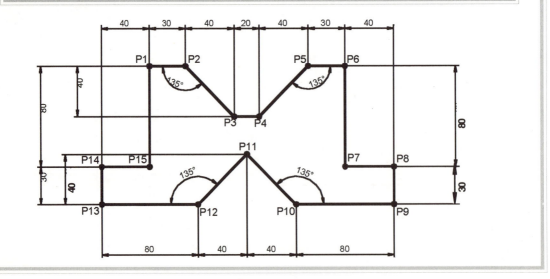

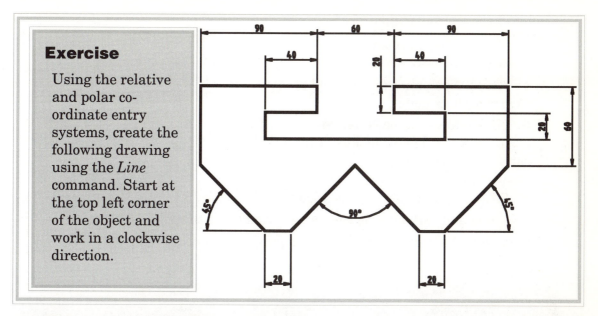

Exercise

Using the relative and polar co-ordinate entry systems, create the following drawing using the *Line* command. Start at the top left corner of the object and work in a clockwise direction.

Saving a drawing file

To Save a drawing, Select *Save As...* from the *File* menu to activate the *Save Drawing As* dialogue box.

Ensure that the location folder you are saving to is correct.

Enter the required file name in the *File name:* edit box then press the *Save* button.

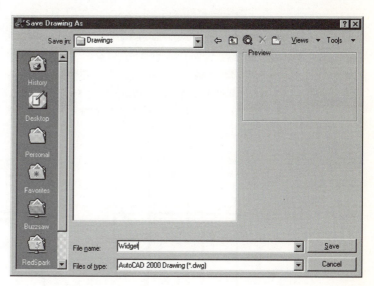

Polyline command

A Polyline is a sequence of connected lines and arcs that are treated as a single object. The Polyline command allows you to switch between straight lines and arcs without exiting the command. You can also control the width of a Polyline.

Draw toolbar	⤶
Draw menu	Polyline
Command line entry	Pline
Alias	PL

2D polylines have the following properties:

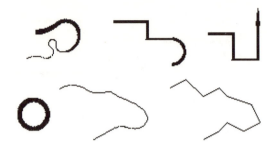

- You can control the width or taper of a polyline,
- Wide 2D polylines can be used to form a circle or donut,
- 2D polylines can be edited to insert, move or delete vertices or to join several lines, arcs and polylines into one polyline,
- fillets and chamfers can be added,
- spline and curve fitting can be performed on a 2D polyline,
- area and perimeter of a closed 2D polyline can be calculated.

Example

From point: *enter* 50,100 ↵ (P1)
Arc/Close/Halfwidth/Length/Undo/Width/<Endpoint of line>: *enter* W *(width)* ↵
Starting width <0.0000>: *enter* 2 ↵
Ending width <2.0000>: *select default* ↵
Arc/Close/Halfwidth/Length/Undo/Width/<Endpoint

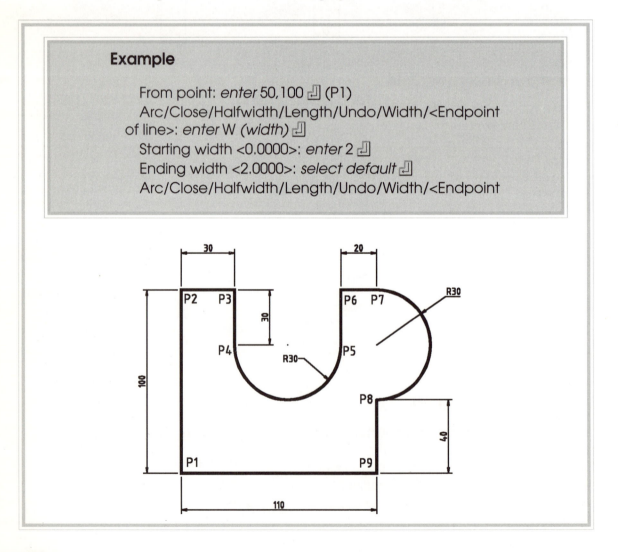

of line>: *enter* @100<90 ⏎ (P2)

    Arc/Close/Halfwidth/Length/Undo/Width/<Endpoint of line>: *enter* @30<0 ⏎ (P3)

    Arc/Close/Halfwidth/Length/Undo/Width/<Endpoint of line>: *enter* @30<270 ⏎ (P4)

    Arc/Close/Halfwidth/Length/Undo/Width/<Endpoint of line>: *enter* A *(Arc)* ⏎

    Angle/CEnter/CLose/Direction/Halfwidth/Line/ Radius/Second pt/Undo/Width/<Endpoint of arc>: *enter* @60<0 ⏎ (P5)

    Angle/CEnter/CLose/Direction/Halfwidth/Line/ Radius/Second pt/undo/width/

    <Endpoint of arc>: *enter* L *(Resume Line segment)* ⏎

    Arc/Close/Halfwidth/Length/Undo/Width/<Endpoint of line>: *enter* @30<90 ⏎ (P6)

    Arc/Close/Halfwidth/Length/Undo/Width/<Endpoint of line>: *enter* @20<0 ⏎ (P7)

    Arc/Close/Halfwidth/Length/Undo/Width/<Endpoint of line>: *enter* A *(Arc)* ⏎

    Angle/CEnter/Close/Direction/Halfwidth/Line/ Radius/Second pt/Undo/Width/<Endpoint of Arc>: *enter* @60<270 ⏎ (P8)

    Angle/CEnter/Close/Direction/Halfwidth/Line/ Radius/Second pt/Undo/Width/

    <Endpoint of Arc>: *enter* L *(Resume Line Segment)* ⏎

    Arc/Close/Halfwidth/Length/Undo/Width/<Endpoint of Line>: *enter* @40<270 ⏎ (P9)

    Arc/Close/Halfwidth/Length/Undo/Width/<Endpoint of Line>: *enter* C *(Close)* ⏎

Once you have created a polyline, you can edit it with the *Pedit* command or use the *Explode* command to convert it to individual line and arc segments. When you explode a polyline to which you have given a width, the line width reverts to 0 and the resulting polyline is positioned along the centre of what was the wide polyline.

More Polyline: the width option

You can use the polyline *width* option to create a variety of shapes.

Example

From point: *pick point* (P1)
Current line-width is 0.0000
Arc/Close/Halfwidth/Length/Undo/Width/<Endpoint of line>: *enter* W ↵
Starting width <0.0000>: *select default* ↵
Ending width <0.0000>: *enter* 2 ↵
Arc/Close/Halfwidth/Length/Undo/Width/<Endpoint of line>: *enter* @40<0 ↵ (P2)
Arc/Close/Halfwidth/Length/Undo/Width/<Endpoint of line>: *enter* W ↵
Starting width <2.0000>: *select default* ↵
Ending width <2.0000>: *enter* 4 ↵
Arc/Close/Halfwidth/Length/Undo/Width/<Endpoint of line>: *enter* @35<270 ↵ (P3)
Arc/Close/Halfwidth/Length/Undo/Width/<Endpoint of line>: *enter* W ↵
Starting width <4.0000>: *select default* ↵
Ending width <4.0000>: *enter* 6 ↵
Arc/Close/Halfwidth/Length/Undo/Width/<Endpoint of line>: *enter* @35<0 ↵ (P4)
Arc/Close/Halfwidth/Length/Undo/Width/<Endpoint of line>: *enter* W ↵
Starting width <6.0000>: *select default* ↵
Ending width <6.0000>: *enter* 8 ↵
Arc/Close/Halfwidth/Length/Undo/Width/<Endpoint of line>: *enter* @35<270 ↵ (P5)
Arc/Close/Halfwidth/Length/Undo/Width/<Endpoint of line>: *enter* W ↵
Starting width <8.0000>: *select default* ↵
Ending width <8.0000>: *enter* 10
Arc/Close/Halfwidth/Length/Undo/Width/<Endpoint of line>: *enter* @75<180 ↵ (P6)
Arc/Close/Halfwidth/Length/Undo/Width/<Endpoint of line>: *enter* W ↵
Starting width <10.0000>: *select default* ↵
Ending width <10.0000>: *enter* 0 ↵
Arc/Close/Halfwidth/Length/Undo/Width/<Endpoint of line>: *enter* C ↵

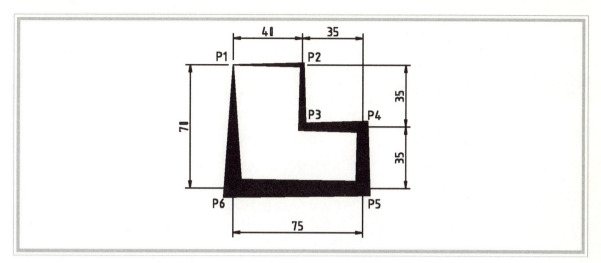

Polyline arc — direction

When drawing a Polyline *Arc*, the arc direction will default in the direction of the previously drawn segment. In order to change the direction of the arc prior to its creation, you will need to use the *Direction* option.

Example

Command: Pline
From point: *pick point* ⏎ (P1)
Current line-width is 0.0000
Arc/Close/Halfwidth/Length/Undo/Width/<Endpoint of line>:*enter* @15<270 ⏎ (P2)

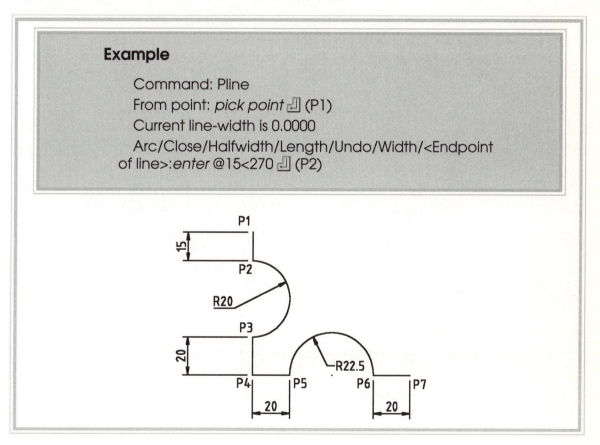

Example continued

Arc/Close/Halfwidth/Length/Undo/Width/<Endpoint of line>: *enter* A ⏎ *(Arc)*

Angle/CEnter/CLose/Direction/Halfwidth/Line/Radius/Second pt/Undo/Width/<Endpoint of arc>: *enter* D ⏎ *(Direction)*

Direction from start point: *enter* @5<0 ⏎ *(indicate direction vector)*

End point: *enter* @40<270 ⏎ *(P3)*

Angle/CEnter/CLose/Direction/Halfwidth/Line/Radius/Second pt/Undo/Width/

<Endpoint of arc>: *enter* L ⏎ *(Return to Line segment)*

Arc/Close/Halfwidth/Length/Undo/Width/<Endpoint of line>: *enter* @20<270 ⏎ *(P4)*

Arc/Close/Halfwidth/Length/Undo/Width/<Endpoint of line>: *enter* @20<0 ⏎ *(P5)*

Arc/Close/Halfwidth/Length/Undo/Width/<Endpoint of line>: *enter* A ⏎ *(Arc)*

Angle/CEnter/CLose/Direction/Halfwidth/Line/Radius/Second pt/Undo/Width/

<Endpoint of arc>: *enter* D ⏎ *(Direction)*

Direction from start point: *enter* @5<90 ø *(indicate direction vector)*

End point: *enter* @45<0 ⏎ *(P6)*

Angle/CEnter/CLose/Direction/Halfwidth/Line/Radius/Second pt/Undo/Width/

<Endpoint of arc>: *enter* **L** ⏎ *(Return to Line segment)*

Arc/Close/Halfwidth/Length/Undo/Width/<Endpoint of line>: *enter* @20<0 ⏎ *(P7)*

Arc/Close/Halfwidth/Length/Undo/Width/of line>: ⏎

Exercises using the polyline command

Use the illustrations as a guide and produce the drawings shown using the Polyline command options.

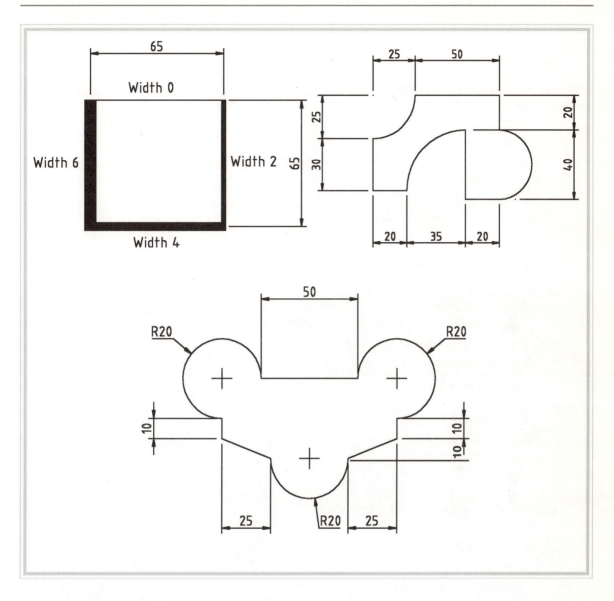

Rectangle command

Draw toolbar	▣
Draw menu	**Rectangle**
Command line entry	**RECTANG**
Alias	**REC**

Used to draw rectangles by designating two points diagonally opposite each other. The rectangle command allows you to reset fillets, chamfers, the rectangle thickness and elevation if required.

Example

Command: rec
Rectang
Chamfer/Elevation/Fillet/Thickness/
Width/<First corner>: *pick any point on the
screen* ↵ (1)
Other corner: *pick another point
diagonally opposite* ↵ (2)

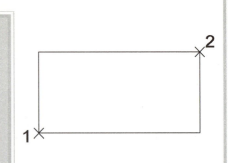

Rectangles with preset fillets

Command: rec
Rectang
Chamfer/Elevation/Fillet/Thickness/
Width/<First corner>: *enter* F ↵
Fillet radius for rectangles <10.0000>:
select default
Chamfer/Elevation/Fillet/Thickness/
Width/<First corner>: *point* 1 ↵
Other corner: *point* 2 ↵

Fillet Rectangle

Rectangles with preset chamfers

Command: rec
Rectang
Chamfer/Elevation/Fillet/Thickness/
Width/<First corner>: *enter* C ↵
First chamfer distance for rectangles
<0.0000>: *enter* 10 ↵
Second chamfer distance for
rectangles <10.0000>: *select default* ↵
Chamfer/Elevation/Fillet/Thickness/
Width/<First corner>: *point* 1 ↵
Other corner: *point* 2 ↵

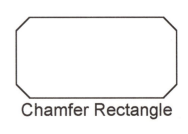

Chamfer Rectangle

Polygon command

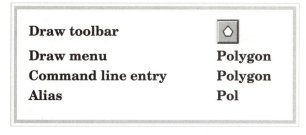

Draw toolbar	⬠
Draw menu	**Polygon**
Command line entry	**Polygon**
Alias	**Pol**

The Polygon command allows you to create an object with number of sides between 3 to 1024.

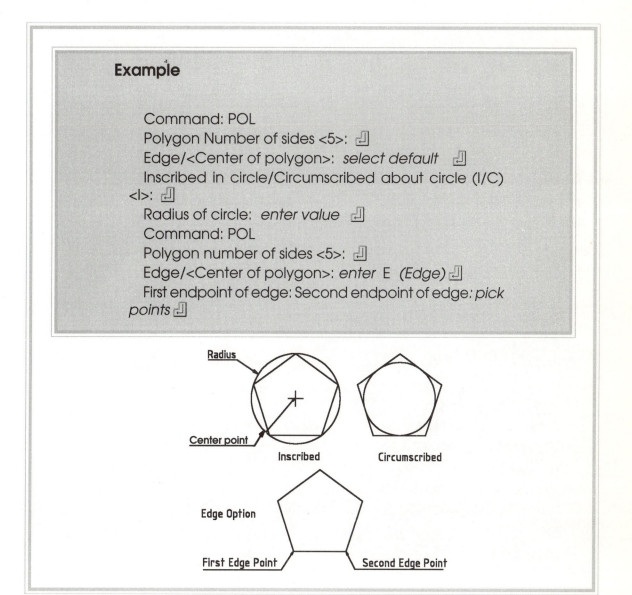

Example

Command: POL
Polygon Number of sides <5>: ↵
Edge/<Center of polygon>: *select default* ↵
Inscribed in circle/Circumscribed about circle (I/C) <I>: ↵
Radius of circle: *enter value* ↵
Command: POL
Polygon number of sides <5>: ↵
Edge/<Center of polygon>: *enter* E *(Edge)* ↵
First endpoint of edge: Second endpoint of edge: *pick points* ↵

Radius

Center point

Inscribed   Circumscribed

Edge Option

First Edge Point   Second Edge Point

Circle command

There are several ways to draw a circle, the default option being the centre, radius option.

Draw toolbar	⊘
Draw menu	**Circle**
Command line entry	**Circle**
Alias	**C**

Example

    Command: c
    Circle 3P/2P/TTR/<Center point>: *enter*
100,100 ⏎
    Circle 3P/2P/TTR/<Center point>:
Diameter/<Radius>: *enter* 50 ⏎

R50.00

Centre point

Exercise

Use the illustrations as a guide and produce a number of circles using each of the options shown. Do not be concerned with dimensional accuracy at this point.

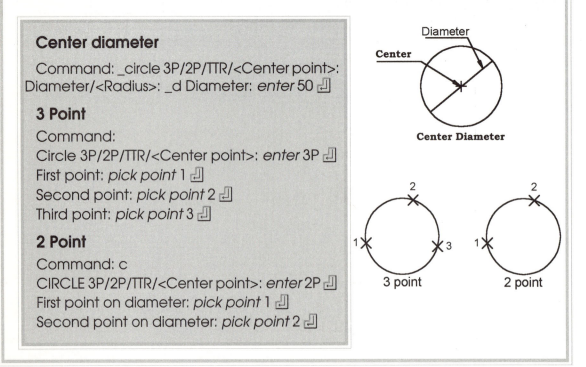

Center diameter

    Command: _circle 3P/2P/TTR/<Center point>:
Diameter/<Radius>: _d Diameter: *enter* 50 ⏎

3 Point

    Command:
    Circle 3P/2P/TTR/<Center point>: *enter* 3P ⏎
    First point: *pick point* 1 ⏎
    Second point: *pick point* 2 ⏎
    Third point: *pick point* 3 ⏎

2 Point

    Command: c
    CIRCLE 3P/2P/TTR/<Center point>: *enter* 2P ⏎
    First point on diameter: *pick point* 1 ⏎
    Second point on diameter: *pick point* 2 ⏎

Diameter

Center

Center Diameter

2
1 3
3 point

2
1
2 point

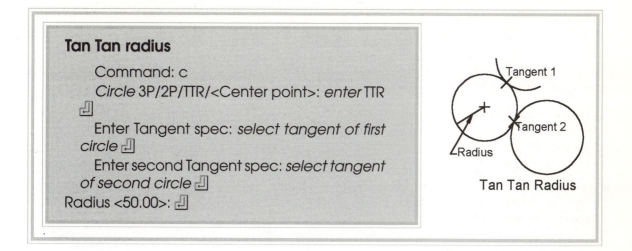

Tan Tan radius

Command: c

Circle 3P/2P/TTR/<Center point>: *enter* TTR ⏎

Enter Tangent spec: *select tangent of first circle* ⏎

Enter second Tangent spec: *select tangent of second circle* ⏎

Radius <50.00>: ⏎

Tan Tan Radius

Exercise — using the circle command

Produce the following drawing using the circle command options.

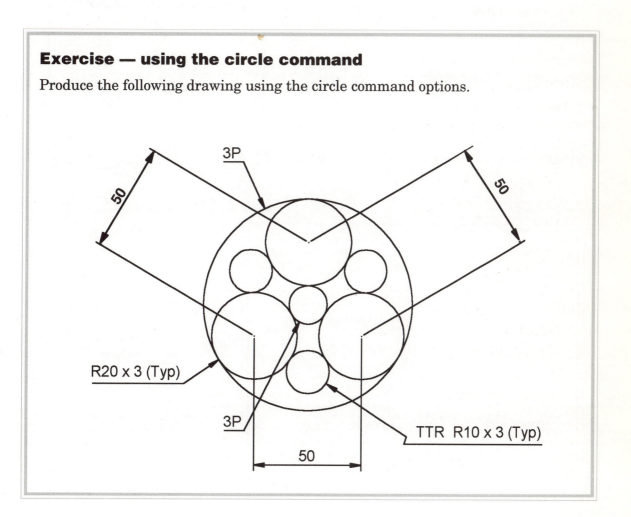

Practical assignment 1

The following assignment will further develop and test your skills using the Line, Polyline and Circle commands. You will need to use the various co-ordinate entry systems covered previously.

The object of Assignment 1 is to construct the basic components of the office swivel chair shown. These items will be further edited and positioned in Assignment 2.

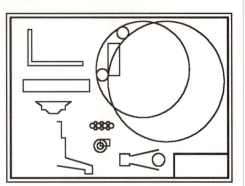

Instructions

Start a *New* drawing using the *Use a Wizard, Quick Setup* option set drawing units to millimetres and set your screen limits from 0,0 to 300,220.
Set your *Grid* and *Snap* to 5.
Place all of the following drawings onto layer 0.

1. Drawing border

Produce the drawing border illustrated below. Set the inner border and title box to a line thickness of 1mm, offset from the outer border by 5mm.

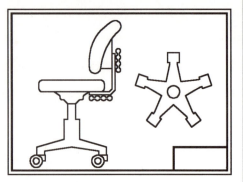

(290,210)

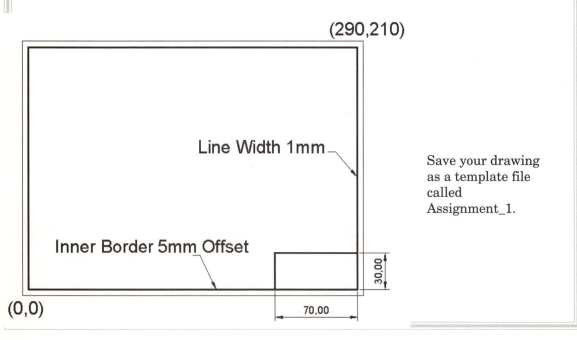

Line Width 1mm

Inner Border 5mm Offset

30,00

70,00

(0,0)

Save your drawing as a template file called Assignment_1.

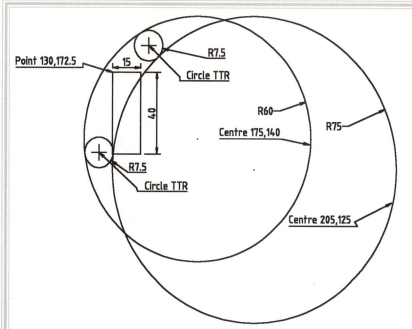

Start a *New* drawing. Select the *Use a Template* button and pick the template file *Assignment_1*.

Save your drawing as Swivel_chair.

2. Chair backrest

Draw the lines and circles shown to the positions indicated.

Use the *TTR* circle option to produce the two 7.5mm radius circles.

Note – rememner to save your work at regular intervals.

3. Backrest bracket
Use the Polyline command to produce the bracket shown.

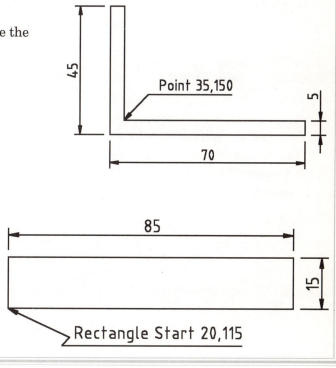

4. Chair seat
Use the Rectangle command to produce the chair seat.

5. Swivel bracket

Use the Polyline command to produce the swivel bracket.

Tip – use the Grid and Snap to locate your co-ordinate positions.

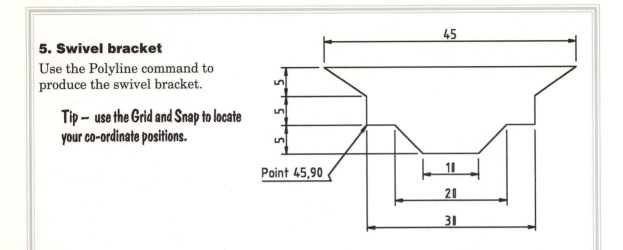

6. Chair base

Use the Polyline command to create the chair base.

You will need to use the polar co-ordinate system to draw the leg feature.

Note – you will create only one half of the base. The completed unit will be generated in Assignment 2 using the Mirror command.

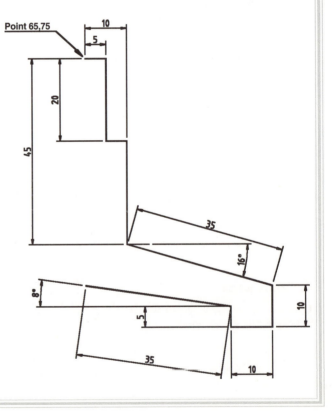

7. Wheel castor

Create the two circles indicated then draw a rectangle located at the circle centres to the dimensions shown.

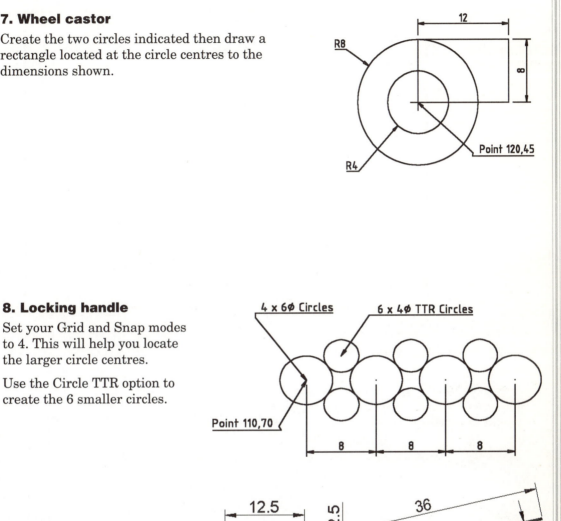

8. Locking handle

Set your Grid and Snap modes to 4. This will help you locate the larger circle centres.

Use the Circle TTR option to create the 6 smaller circles.

9. Base plan

Use both the relative and polar co-ordinate systems to create the polyline shown.

Tip – reset grid and snap to 2.5.

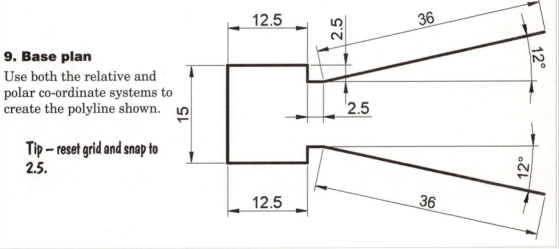

Your drawing
should now
look like this
illustration.

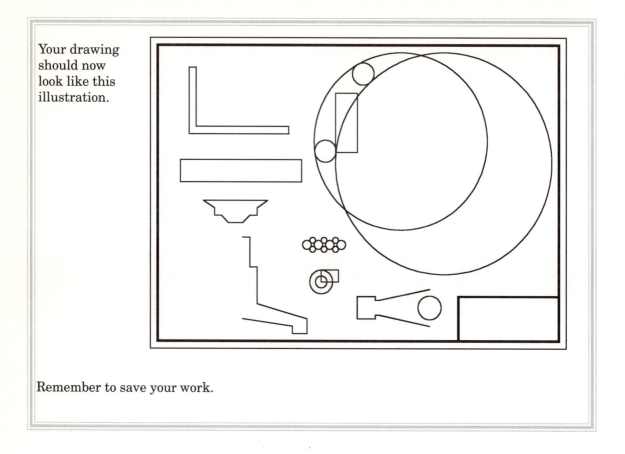

Remember to save your work.

Multiple choice questions

1   Which of the following answers gives the Absolute position of B?

A   20,30

B   20,10

C   40,20

D   40,40

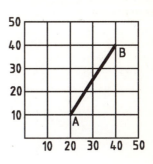

2   Which of the following answers gives point B as a Relative co-ordinate from point A?

A   @20,30

B   @40,30

C   @30,20

D   @30,40

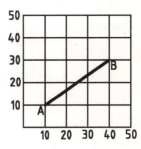

3   Which of the following answers gives point B as a Polar co-ordinate from point A?

A   @30<180

B   @63.64<180

C   @63.64<135

D   @135<63.64

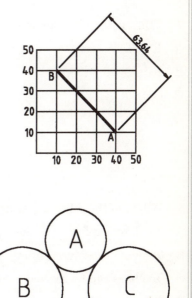

4   Given that circles B and C already exist, which is the most efficient circle option to create circle A?

A   Tan Tan Radius

B   3 Point

C   2 Point

D   Diameter

5   Which of the following is a fillet?

A

B

C

D

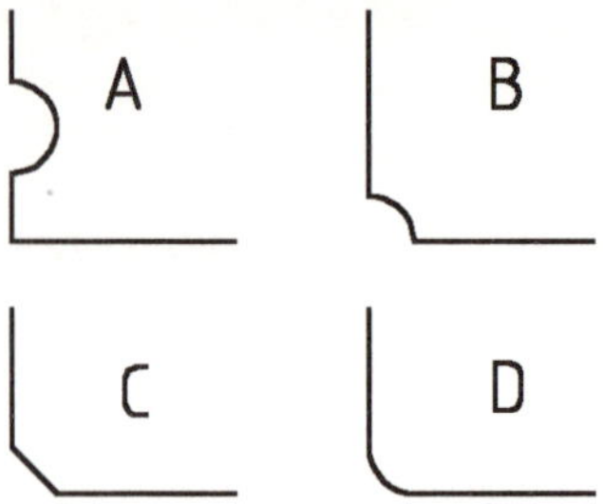

6   Which command and option would you use to create the following object?

A   Line, Thickness

B   Line, Point

C   Polyline, Direction

D   Polyline, Width

7   What is the alias for the Rectangle command?

A   RECT

B   REC

C   RE

D   R

8   Which of the following statements is not true?

A   Toolbars cannot be docked at the screen sides

B   Toolbars can placed in any location

C   Toolbars can be resized

D   Selecting a Toolbar that has a black triangle in the corner will result in a Flyout toolbar appearing

Chapter 3

Drawing aids

AutoCAD provides a range of drawing aids that can be used to improve drawing accuracy and speed. The features covered in this chapter include the Grid, Snap, Ortho and Polar modes, which help automate object positioning to either exact co-ordinate locations or by forcing cursor movement at specific pre-determined angles.

Objectives

At the end of this chapter you will be able to:

▷   Set Grid and Snap values.

▷   Set rectangular and Isometric Snap.

▷   Use function keys to toggle modes on/off.

▷   Create a drawing using the Polar Tracking mode.

▷   Understand and use the Grid/Snap/Polar Setting dialogue box.

▷   Understand and use object snap modes.

New commands

▷   Osnap

▷   Polar Snap

▷   Ellipse   EL

Grid, Snap, Ortho and Polar modes

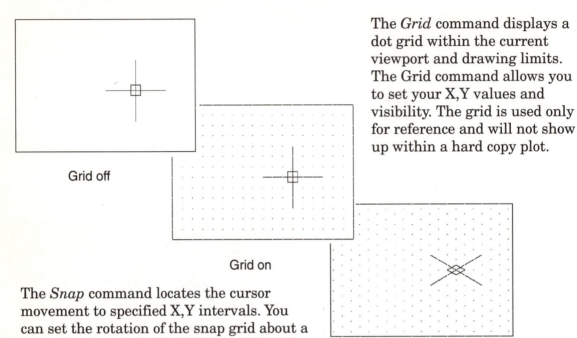

Grid off

Grid on

The *Grid* command displays a dot grid within the current viewport and drawing limits. The Grid command allows you to set your X,Y values and visibility. The grid is used only for reference and will not show up within a hard copy plot.

The *Snap* command locates the cursor movement to specified X,Y intervals. You can set the rotation of the snap grid about a base point, set individual X and Y spacing values, or choose an isometric format for the snap grid. The Snap grid is invisible but you can set it to equal the Grid X and Y values.

Ortho mode forces the movement of the cursor in horizontal or vertical directions only.

You can set Ortho mode, Grid and Snap values by using the *Drafting Settings* dialogue box which can be activated by selecting *Drafting Setting...* located within the *Tools* pull-down menu.

Grid, Snap, Ortho and Polar Tracking can also be activated by pressing their respective buttons on the Status bar or by using the following function keys:

F7   Toggle Grid On/Off
F8   Toggle Ortho mode On/Off
F9   Toggle Snap On/Off
F10 Toggle Polar Tracking On/Off

Example – using grid and snap

1. Start a *New* drawing using the *Start from Scratch – Metric* option. Select *Drafting Settings...* from the *Tools* menu and ensure that the *Snap and Grid* tab is active.
2. Set the Snap X and Y and the Grid X and Y spacing values to 10. Activate the Snap On and Grid On tick boxes.

Snap and Grid	Polar Tracking	Object Snap

☑ Snap On (F9)   ☑ Grid On (F7)

┌─ Snap ─────────────────────┐ ┌─ Grid ─────────────────────┐
│ Snap X spacing: | 10 | │ │ Grid X spacing: | 10 | │
│ Snap Y spacing: | 10 | │ │ Grid Y spacing: | 10 | │
└────────────────────────────┘ └────────────────────────────┘

3. Move the cursor about the screen and you will see the cursor *Snap* to the Grid. You can set different values for the *Snap* and the *Grid*.
4. Activate the *Drafting Settings* dialogue box and enter a Snap X and Y value of 5.

☑ Snap On (F9)

┌─ Snap ─────────────────────┐
│ Snap X spacing: | 5 | │
│ Snap Y spacing: | 5 | │
│ Angle: | 45 | │
└────────────────────────────┘

5. Move the cursor about the screen again and you will see that the cursor is *Snapping* to the spacing between the Grid as well as to the Grid co-ordinates.
6. Activate the *Drafting Settings* dialogue box and set an Angle value of 45. Notice how the *Snap* and *Grid* have been rotated through 45 degrees.

7.   Reset the Snap X and Y spacing to 10 and the *Angle* value to 0 in the *Drafting Settings* dialogue box.

8.   Activate the *Isometric snap* radio button under *Snap type & style* then press *OK*. The Snap and Grid now adopt an Isometric pattern.

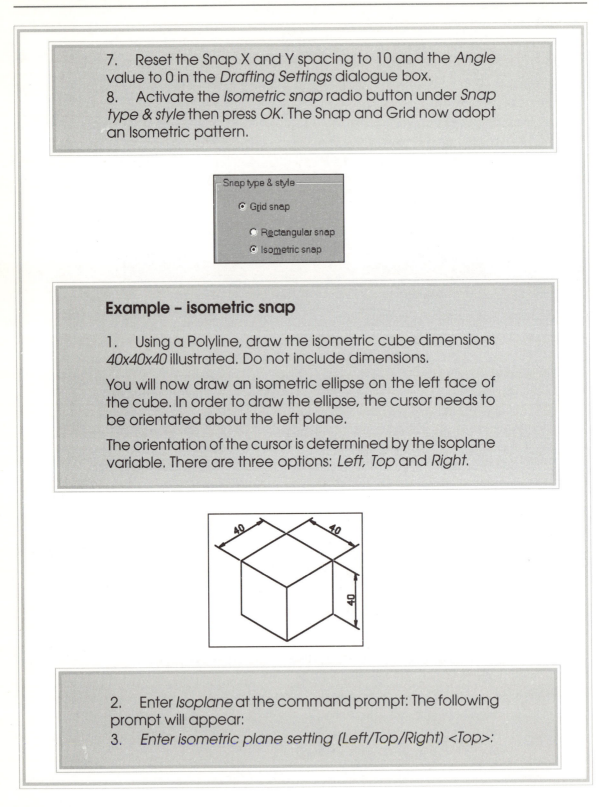

Example – isometric snap

1.   Using a Polyline, draw the isometric cube dimensions *40x40x40* illustrated. Do not include dimensions.

You will now draw an isometric ellipse on the left face of the cube. In order to draw the ellipse, the cursor needs to be orientated about the left plane.

The orientation of the cursor is determined by the Isoplane variable. There are three options: *Left, Top* and *Right*.

2.   Enter *Isoplane* at the command prompt: The following prompt will appear:

3.   *Enter isometric plane setting (Left/Top/Right) <Top>:*

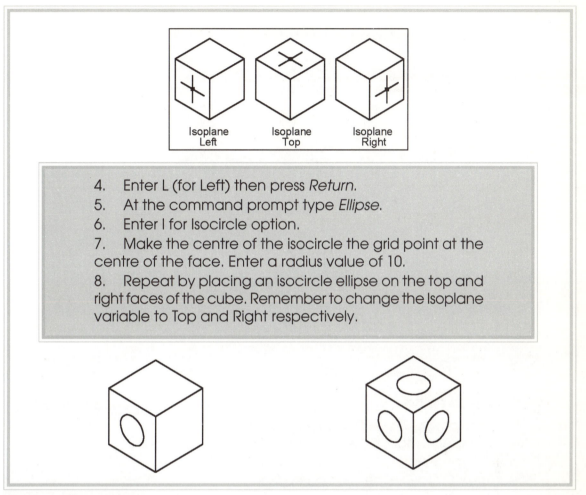

Isoplane Left | Isoplane Top | Isoplane Right

4.	Enter L (for Left) then press *Return*.
5.	At the command prompt type *Ellipse*.
6.	Enter I for Isocircle option.
7.	Make the centre of the isocircle the grid point at the centre of the face. Enter a radius value of 10.
8.	Repeat by placing an isocircle ellipse on the top and right faces of the cube. Remember to change the Isoplane variable to Top and Right respectively.

Polar Tracking

When the Polar Tracking mode is active movement of the cursor at pre-selected angles will trigger a polar direction track.

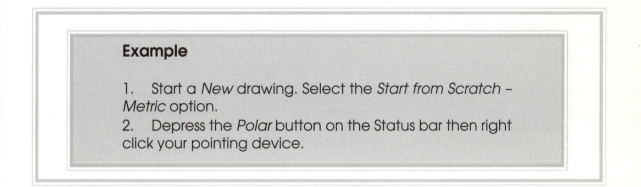

Example

1.	Start a *New* drawing. Select the *Start from Scratch – Metric* option.
2.	Depress the *Polar* button on the Status bar then right click your pointing device.

Drafting Settings

Snap and Grid | Polar Tracking | Object Snap

☑ Polar Tracking On (F10)

Polar Angle Settings

Increment angle:

30

☐ Additional angles

New

Delete

Object Snap Tracking Settings

○ Track orthogonally only

● Track using all polar angle settings

Polar Angle measurement

● Absolute

○ Relative to last segment

Options...   OK   Cancel   Help

Press *Setting…*

This will activate the *Drafting Settings… Polar Tracking* tab.

3. Activate the *Polar Tracking On* check box, set an incremental angle of 30 and check the *Track using all polar angle settings* radio button.

4. Press *OK* to exit the dialogue box.

5. Activate the Line command.

6. Start the line from co-ordinate 50,50 ↵.

7. Move the cursor vertically above the start point. A dotted tracking line will appear with a tooltip indicating the current distance and direction from your start point. Enter 50 ↵.

8. Move the cursor approximately 30 degrees above the horizontal to the right of the line. When the dotted tracking line appears enter a direct distance of 50 ↵.

On
Off
Settings...

POLAR

Polar: 27.5871 < 90°

Polar: 45.7420 < 30°

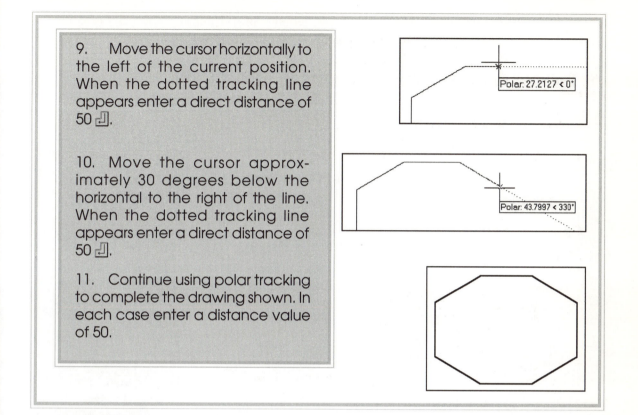

9. Move the cursor horizontally to the left of the current position. When the dotted tracking line appears enter a direct distance of 50 ⏎.

10. Move the cursor approximately 30 degrees below the horizontal to the right of the line. When the dotted tracking line appears enter a direct distance of 50 ⏎.

11. Continue using polar tracking to complete the drawing shown. In each case enter a distance value of 50.

Object Snap

Another major drawing aid is the Object Snap command. Object Snap lets you snap to specific geometric points of existing objects in the drawing. For instance, you may wish to end a line to the middle of an existing line, in which case you would snap to the mid-point of the existing line. You can specify a variety of Object Snap modes.

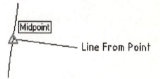

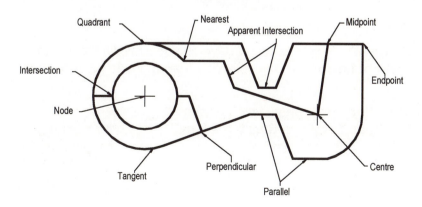

This diagram illustrates the modes that are available for object snap:

Object Snap modes can be activated from the Object Snap toolbar. When the graphics cursor is passed over the toolbar, the cursor will change into an arrow. As this arrow passes over each of the tools, the name of the tool will be displayed. The following figure illustrates the name and position of each of the Object Snap modes in the toolbar.

Temporary Tracking Point

Snap From

Endpoint

Midpoint

Intersection

Apparent Intersection

Extension

Centre

Quadrant

Tangent

Perpendicular

Parallel

Swap to Insert

Mode

Nearest

None

Osnap Settings

You can apply an Object Snap mode in any one of three ways:

● By selecting the appropriate toolbar button.

● By holding down the Shift Key and pressing the right mouse button to reveal the dialogue box to the right.

● By typing in the first three letters of the snap mode at the command line.

Temporary track point

From

Point Filters

Endpoint

Midpoint

Intersection

Apparent Intersect

Extension

Center

Quadrant

Tangent

Perpendicular

Parallel

Node

Insert

Nearest

None

Osnap Settings...

To select a point using an object snap, position the cursor over the desired object. The AutoSnap (see below) feature will cause an icon to appear depending on which object snap option(s) is active. For instance, if *Endpoint* is active, then a little square will appear at the end of the line closest to the cursor. Similarly, if *Centre* is active then a circle will appear at the centre of any circle or arc that the cursor passes. Only objects that are visible on the screen will be chosen.

Running Object Snap

The Running Object Snap feature allows you to keep active one or more Object Snap modes. This saves you the bother of having to keep selecting a snap mode prior to each point selection.

You can activate Running Object Snap by typing *Osnap* at the command line or selecting the Osnap button on the status bar then right click the pointing device.

Press *Settings...* to activate the *Drafting Settings... Object Snap* tab.

52

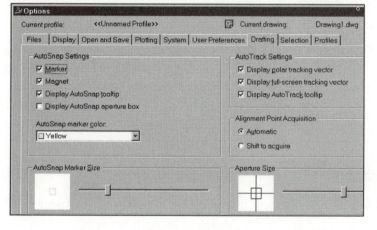

Check the required snap mode tick boxes that you wish to remain active.

AutoSnap

AutoSnap is used to activate the Marker, Magnet, Tooltip and Aperture box settings which enable a marker to be shown as each Object Snap mode is activated. AutoSnap Settings are located within the *Drafting* tab in the *Options* dialogue box.

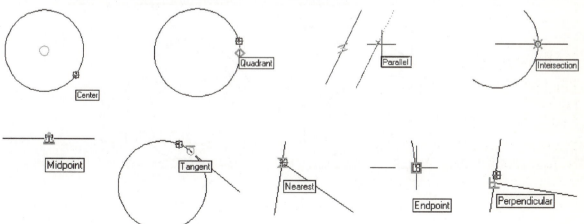

Exercises using the object snap feature

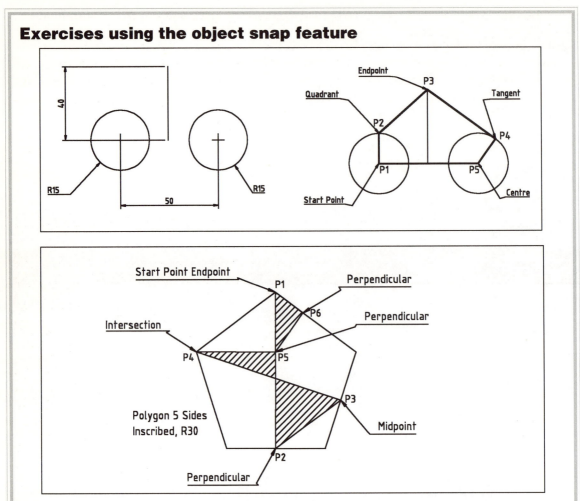

Use only the Intersection and Quadrant *Osnaps* in the following Running Object Snap exercise.

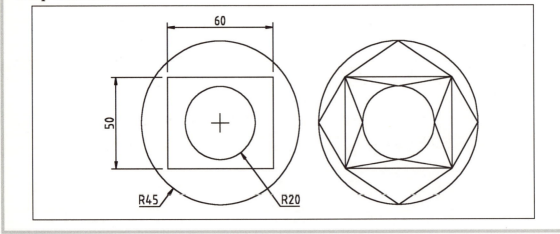

Customising toolbars and keyboard shortcuts

Whilst toolbars present a quick and efficient method for activating commands, too many toolbars on the graphics screen can be confusing and take up valuable space. An alternative to having too many toolbars visible in the graphics window is to create a customised toolbar containing only those commands that you would require on a regular basis.

For instance, a single toolbar could contain just a selection of the most used drawing and editing commands.

The Customise dialogue box allows you to alter existing toolbars and create new ones.

Activate the Customise dialogue box by selecting *Toolbars...* from the View drop down menu. Alternatively, right-click any tool icon then select *Customise...* at the bottom of the list.

The Toolbars window illustrates the number and type of toolbars that are available. A tick indicates that the toolbar is currently active.

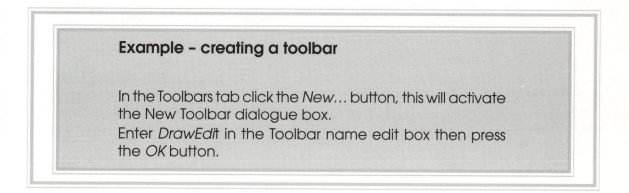

The following example will illustrate how to create a new customised toolbar.

Example – creating a toolbar

In the Toolbars tab click the *New...* button, this will activate the New Toolbar dialogue box.
Enter *DrawEdit* in the Toolbar name edit box then press the *OK* button.

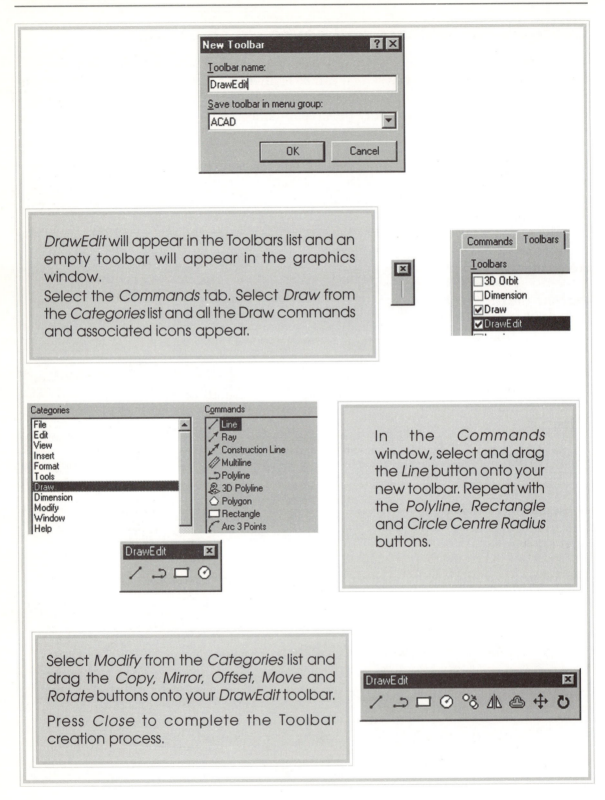

New Toolbar

Toolbar name:
DrawEdit

Save toolbar in menu group:
ACAD

OK   Cancel

DrawEdit will appear in the Toolbars list and an empty toolbar will appear in the graphics window.

Select the *Commands* tab. Select *Draw* from the *Categories* list and all the Draw commands and associated icons appear.

Commands Toolbars

Toolbars
☐ 3D Orbit
☐ Dimension
☑ Draw
☑ DrawEdit

Categories
File
Edit
View
Insert
Format
Tools
Draw
Dimension
Modify
Window
Help

Commands
Line
Ray
Construction Line
Multiline
Polyline
3D Polyline
Polygon
Rectangle
Arc 3 Points

DrawEdit

In the *Commands* window, select and drag the *Line* button onto your new toolbar. Repeat with the *Polyline, Rectangle* and *Circle Centre Radius* buttons.

Select *Modify* from the *Categories* list and drag the *Copy, Mirror, Offset, Move* and *Rotate* buttons onto your *DrawEdit* toolbar.

Press *Close* to complete the Toolbar creation process.

DrawEdit

Creating new tool buttons

Not every AutoCAD command has an associated tool button, therefore there may be times when you will need to create your own tool buttons to include within your customised toolbar.

Commands | Toolbars | Button Properties | Keyboard

Name: Line

Description: Creates straight line segments: LINE

Button Image

Edit...

Macro associated with this button:
^C^C_line

Every tool button has a number of properties. Activate the Customise dialogue box.

Press the Line button within the *DrawEdit* toolbar. This will activate the *Button Properties* tab.

You can see that the Line Button has a name, description, macro and image attached to it.

The following example will illustrate how to create a customised user defined button.

Example – creating a new button

Press the *Commands* tab and select User Defined from the *Categories* list and drag the User Defined Button from the *Commands* window onto the *DrawEdit* toolbar.
Press the User Defined Button in the toolbar to activate its default button properties.

Note: the button may appear invisible in the toolbar. Don't worry, it is there!

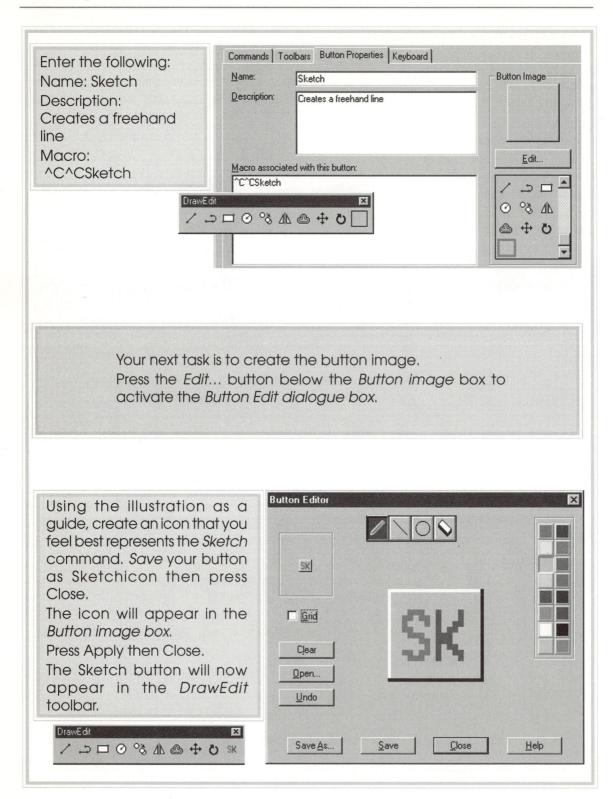

Enter the following:
Name: Sketch
Description:
Creates a freehand line
Macro:
^C^CSketch

Your next task is to create the button image.
Press the *Edit...* button below the *Button image* box to activate the *Button Edit dialogue box*.

Using the illustration as a guide, create an icon that you feel best represents the *Sketch* command. *Save* your button as Sketchicon then press Close.
The icon will appear in the *Button image box*.
Press Apply then Close.
The Sketch button will now appear in the *DrawEdit* toolbar.

Keyboard shortcuts

Experienced AutoCAD users may find that using the keyboard to activate a command is faster than using the menus or toolbars.

Most commands have a shortcut alias which when typed at the command line will activate the command, for instance, typing M will activate the Move command.

You can assign a shortcut key by using the Keyboard tab.

Example – Assigning a Keyboard Shortcut

Press the Keyboard tab in the *Customise* dialogue box.

Select Draw Menu from the *Categories* list.

Scroll down the *Draw* commands and select *Donut.*

Customize dialog:
Tabs: Commands | Toolbars | Properties | Keyboard

Categories: Draw Menu
Commands: Circle Tan Tan Radius, Circle Tan Tan Tan, Donut, Spline, Ellipse Center, Ellipse Axis End, Ellipse Arc, Block Make, Block Base

Menu Group: ACAD
Current Keys: (empty)
Press new shortcut key: Ctrl+D
Buttons: Assign, Remove, Show All...

Description: Creates filled circles and rings: DONUT
Currently assigned to: Toggles coordinate display

Buttons: Close, Help

Click into the *Press New Shortcut Key* field then press the CTRL and D keys.

Press the Assign button and *CTRL+D* will appear in the *Current Keys* window.

Press Close to exit the dialogue box.

CTRL D will now activate the *Donut* command.

Chapter 4

Basic editing tools

The object of this chapter is to introduce you to a range of AutoCAD editing commands. A summary of the editing commands is included to give an overview of the editing commands features prior to their use. Exercises are included throughout this chapter and conclude with a practical assignment to test your understanding of the features covered.

Objective

At the end of this chapter you will be able to:

▷ Use a variety of editing commands shown below.

New commands

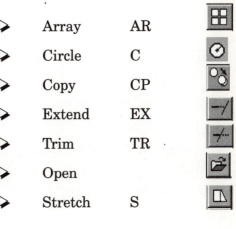

▷	Mirror	MI		▷	Array	AR
▷	Move	M		▷	Circle	C
▷	Erase	E		▷	Copy	CP
▷	Offset	OF		▷	Extend	EX
▷	Rotate	RO		▷	Trim	TR
▷	Fillet	F		▷	Open	
▷	Chamfer	CHA		▷	Stretch	S

Object selection

During the execution of many of the editing commands you will be prompted to select objects. For instance, if you need to move a number of objects about the screen you would activate the *Move* command and you will be prompted to select the objects you require to move.

There are a number of ways in which you can select a single object or multiple objects when the selection of objects are required to be processed. After the editing command has been activated, AutoCAD prompts: *Select objects*.

You can select objects by typing keywords and/or by using the cursor. Below are some of the options available.

Window (W) – Lets you designate all objects that fall completely within a window.

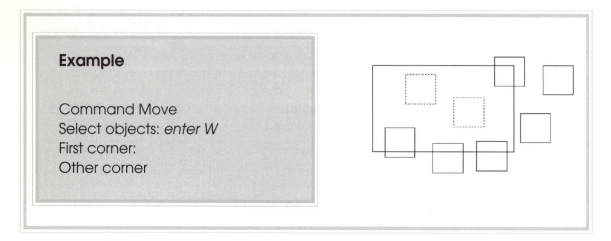

Example

Command Move
Select objects: *enter W*
First corner:
Other corner

Crossing (C) – Similar to window except that objects contained within the rectangular boundary or crossing will be selected

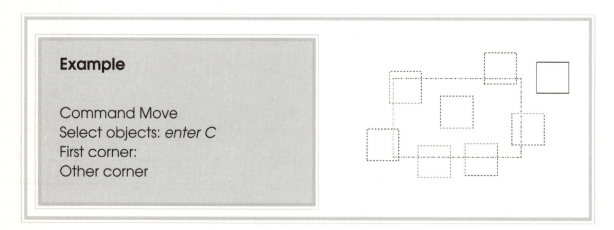

Example

Command Move
Select objects: *enter C*
First corner:
Other corner

Fence (F) – Selects all objects crossing a selection fence. The fence can be any shape and can be open ended.

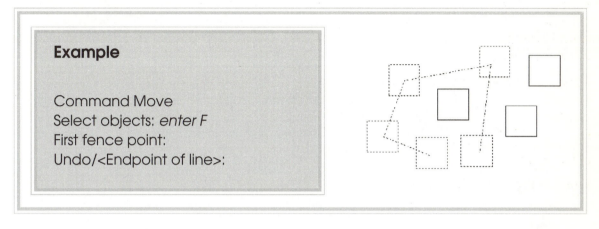

Example

Command Move
Select objects: *enter F*
First fence point:
Undo/<Endpoint of line>:

All – Selects all objects in the drawing, including those on frozen and locked layers.

Last (L) – Selects the most recently created object.

Previous (P) – Selects the most recent selection set.

You will use most of these selection options throughout the course of this book.

Erase command

Modify toolbar	
Modify menu	**Erase**
Command line entry	**Erase**
Alias	**E**

The Erase command is used to permanently erase unwanted objects.

Create a close approximation of the following drawing using the Rectangle and Circle commands.

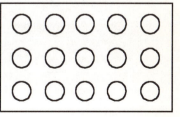

We can use the Erase command to demonstrate the various object selection methods covered previously.

Example

Activate the Erase command

Command: erase

(Use the screen cursor to individually select the two left side columns of circles)

Select objects:	1	found	↵
Select objects:	1	found	↵
Select objects:	1	found	↵
Select objects:	1	found	↵
Select objects:	1	found	↵
Select objects:	1	found	↵

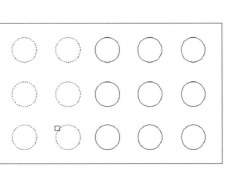

After each example, at the command prompt, enter *Undo* to restore your drawing.

Example

Using the *Window* selection method to select only those objects completely within the window.

Command: erase

Select objects: *enter W* (Window) ↵

First corner: Other corner: 3 found ↵

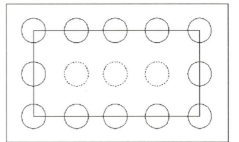

Example

Using the *Crossing* selection method to select those objects completely within the window and touching the window.

Command: erase

Select objects: *enter C* (Crossing) ↵

First corner: Other corner: 15 found ↵

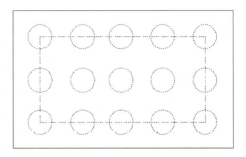

Example

Using the *Fence* selection method to select only those objects that the fence line touches.

Command: erase

Select objects: *enter F* (Fence) ⏎

First fence point: *Follow the path indicated* Undo/<Endpoint of line>: 9 found ⏎

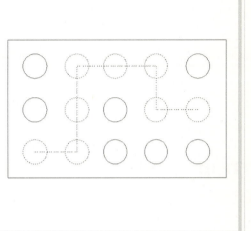

Extend command

Modify toolbar	
Modify menu	**Extend**
Command line entry	**Extend**
Alias	**EX**

Used to extend existing objects so that they end precisely at a boundary edge as defined by another object or a projected boundary.

Create a close approximation of the following drawing using the *Line* command.

Example

Activate the *Extend* command

Extend

Select boundary edges: (Projmode = UCS, Edgemode = No extend)

Select objects: *use a Crossing box to select all the lines*

Select objects: Other corner: 7 found

<Select object to extend>/Project/ Edge/Undo: *select the 6 lines either side of the vertical line*

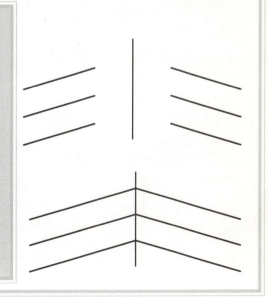

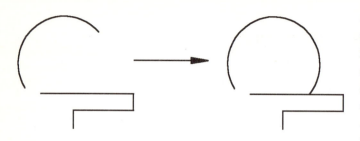

An object will extend along its original path to another object, for instance, in the example shown, given that both the arc and line are selected as boundary edges, then the arc will extend along its original path to its intersection point with the line.

Trim command

Modify toolbar	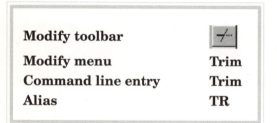
Modify menu	**Trim**
Command line entry	**Trim**
Alias	**TR**

Used to trim objects so they end precisely at a cutting edge defined by another object.

Create a close approximation of the following drawing using the *Line* command.

Example

Activate the *Trim* command

Command: _trim

Select cutting edges:(Projmode = UCS, Edgemode = No extend)

Select objects: *use a Crossing box to select all the lines* Other corner: 6 found ⏎

<Select object to trim>/Project/Edge/Undo: *pick the ends of the 4 near-horizontal lines* ⏎

<Select object to trim>/Project/Edge/Undo: ⏎

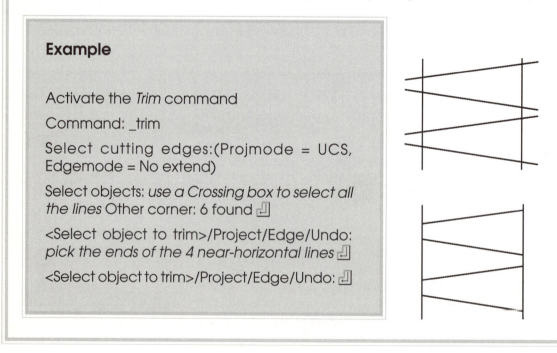

Exercises — using the trim command

Attempt the following Trim exercises.

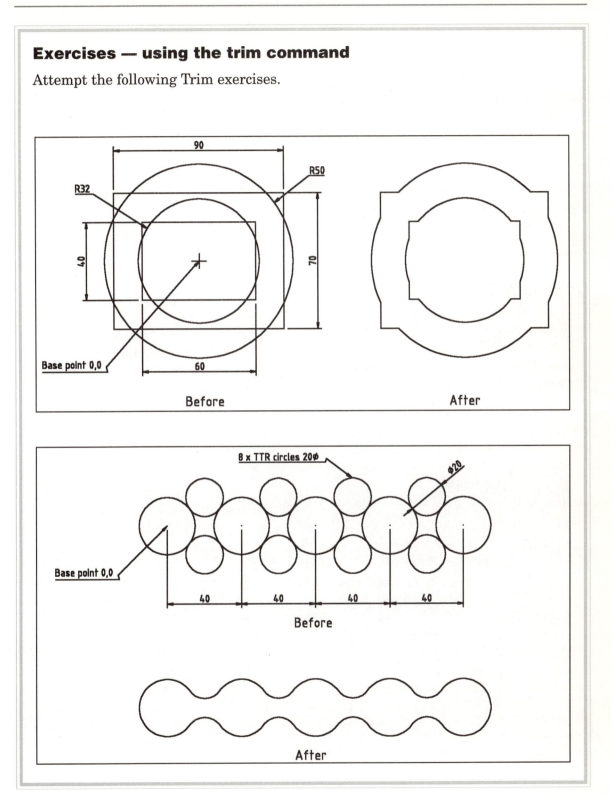

Base point 0,0

Before

After

8 x TTR circles 20⌀

⌀20

Base point 0,0

Before

After

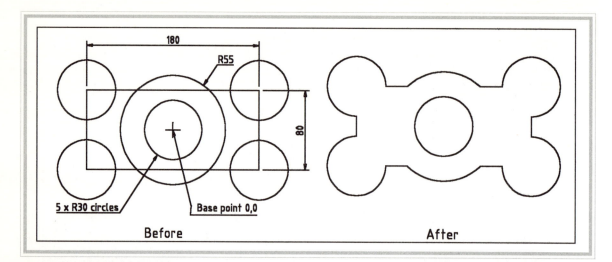

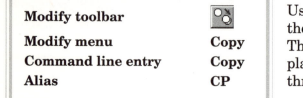

Copy command

Modify toolbar	
Modify menu	**Copy**
Command line entry	**Copy**
Alias	**CP**

Used to copy an existing object and place the copied object at a specified location. The Copy *Multiple* option allows you to place multiple copies of selected objects throughout the drawing.

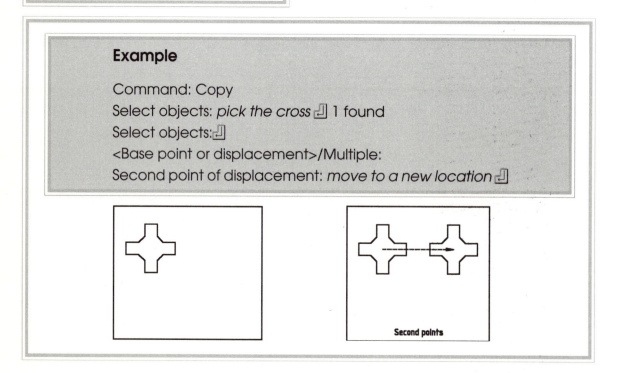

Example

Command: Copy
Select objects: *pick the cross* ⏎ 1 found
Select objects: ⏎
<Base point or displacement>/Multiple:
Second point of displacement: *move to a new location* ⏎

Second points

Multiple copy example

Command: Copy
Select objects: *pick the cross* ↵ 1 found
Select objects: ↵
<Base point or displacement>/Multiple: *enter M*
Base point: Second point of displacement:
Second point of displacement: *pick point* ↵
Second point of displacement: *pick point* ↵
Second point of displacement: *pick point* ↵
Second point of displacement: *pick point* ↵

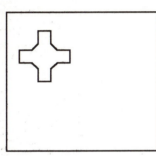

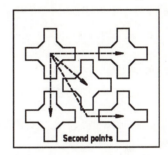

Second points

Note – copied objects can also be located using Direct Distance entry.

Rotate command

Modify toolbar	↻
Modify menu	**Rotate**
Command line entry	**Rotate**
Alias	**RO**

Used to change the orientation of existing objects by rotating them about a fixed base point. A positive rotation angle rotates the object(s) anti-clockwise, whilst a negative rotation angle rotates the object(s) clockwise.

Example

Command: RO
Rotate
Select objects: *pick the arrow* 1 found
Select objects:
Base point: *pick the centre of the arrow*
<Rotation angle>/Reference: *enter 45*

Command: RO
Rotate
Select objects: *pick the arrow* 1 found
Select objects:
Base point: *pick the centre of the arrow*
<Rotation angle>/Reference: *enter –45*

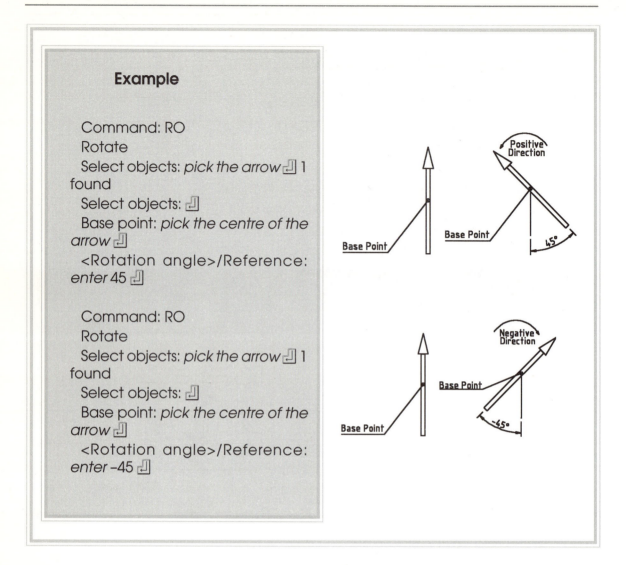

Offset command

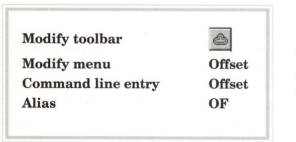

Modify toolbar	
Modify menu	Offset
Command line entry	Offset
Alias	OF

The *Offset* command creates a new object at a specified distance from an existing object or through a specified point.

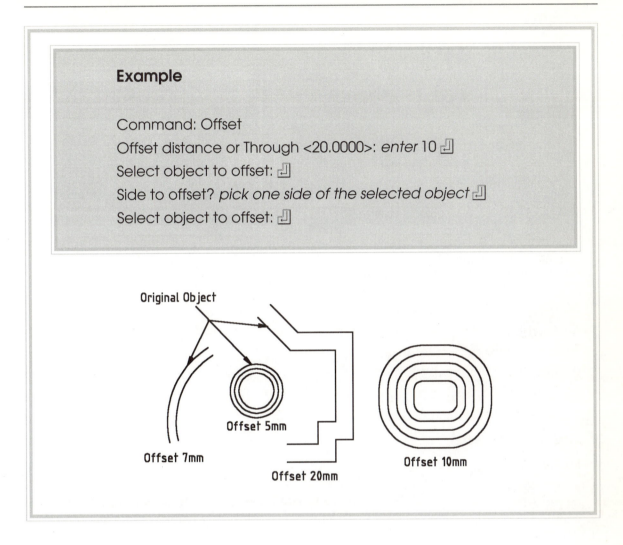

Example

Command: Offset

Offset distance or Through <20.0000>: *enter* 10 ⏎

Select object to offset: ⏎

Side to offset? *pick one side of the selected object* ⏎

Select object to offset: ⏎

Original Object

Offset 5mm

Offset 7mm

Offset 20mm

Offset 10mm

Stretch command

Modify toolbar	⬜
Modify menu	Stretch
Command line entry	Stretch
Alias	S

Lets you move or stretch objects that cross the selection window. The *Stretch* command also moves vertices of 2D solids that lie inside the window and leaves those outside unchanged.

Note – objects to be stretched must be selected using a crossing box.

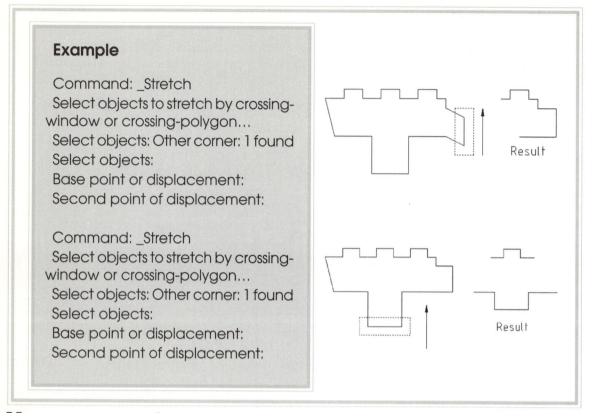

Example

Command: _Stretch
Select objects to stretch by crossing-window or crossing-polygon...
Select objects: Other corner: 1 found
Select objects:
Base point or displacement:
Second point of displacement:

Command: _Stretch
Select objects to stretch by crossing-window or crossing-polygon...
Select objects: Other corner: 1 found
Select objects:
Base point or displacement:
Second point of displacement:

Move command

Modify toolbar	
Modify menu	**Move**
Command line entry	**Move**
Alias	**M**

Lets you move one or more objects from one location to another without changing orientation.

Example

Command: Move
Select objects: *pick the cross* ⏎
Select objects: ⏎
Base point or displacement: *pick the centre of the cross* ⏎
Second point of displacement: *enter @50<0* ⏎

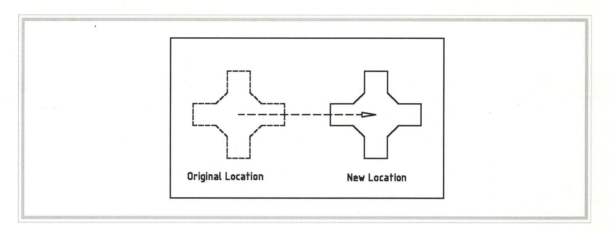

Fillet command

Modify toolbar	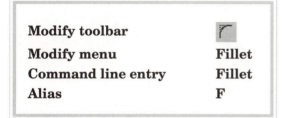
Modify menu	**Fillet**
Command line entry	**Fillet**
Alias	**F**

Used to connect two lines, arcs or circles by means of a smoothly fitted arc of a specified radius. The fillet command can be applied to an entire polyline if the *Fillet-Polyline* option is selected.

Radius option

Example

Command: Fillet

(*Trim* mode) Current fillet radius = 0.0000

Polyline/Radius/Trim/<Select first object>: *R*

Enter fillet radius <0.0000>: *enter* 5 ⏎

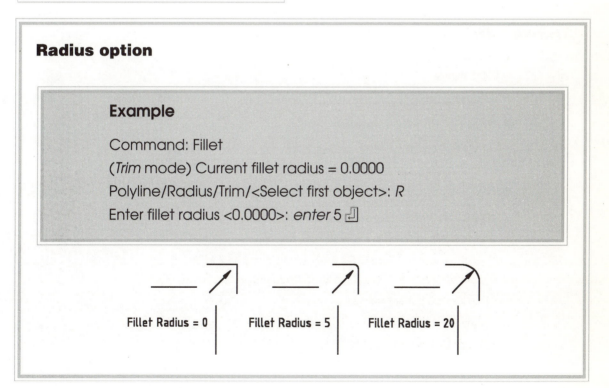

Polyline option

Example

Command: Fillet

(*Trim* mode) Current fillet radius = 5.0000

Polyline/Radius/Trim/<Select first object>: enter *P* ⏎

Select 2D polyline: *pick the polyline* ⏎

12 lines were filleted

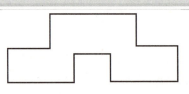

Fillet Polyline
Option

Trim option

Example

Command: Fillet

(*Trim* mode) Current fillet radius = 5.0000

Polyline/Radius/Trim/<Select first object>: *enter T* ⏎

Trim/No trim <Trim>: *enter N* ⏎

Polyline/Radius/Trim/<Select first object>: *pick line* ⏎

Select second object: *pick line* ⏎

Command: Fillet

(*Notrim* mode) Current fillet radius = 5.0000

Polyline/Radius/Trim/<Select first object>: *enter T* ⏎

Trim/No trim<No trim>: *enter T* ⏎

Polyline/Radius/Trim/<Select first object>: *pick line* ⏎

Select second object: *pick line* ⏎

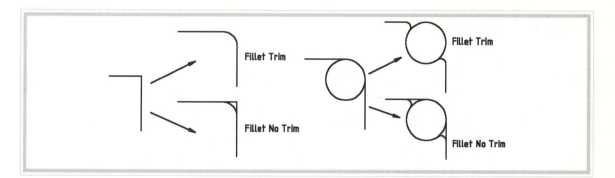

Chamfer command

Modify toolbar	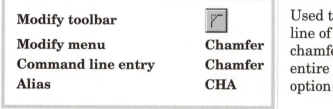
Modify menu	**Chamfer**
Command line entry	**Chamfer**
Alias	**CHA**

Used to connect two lines by means of a line of a specified length or angle. The chamfer command can be applied to an entire polyline if the *Chamfer-Polyline* option is selected.

Polyline option

Example

Command: Chamfer
(*Trim* mode) Current chamfer Dist1 = 5.0000, Dist2 = 5.0000
Polyline/Distance/Angle/Trim/Method/<Select first line>: *enter P*
Select 2D polyline: *select the polyline*
12 lines were chamfered

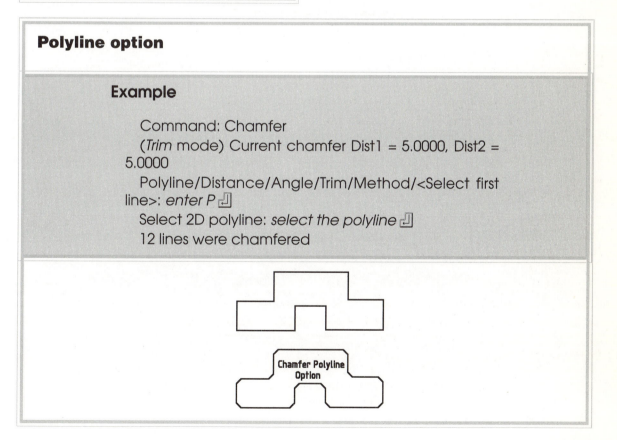

Chamfer Polyline
Option

Distance option

Example

Command: Chamfer
(*Trim* mode) Current chamfer Dist1 = 10.0000, Dist2 = 5.0000
Polyline/Distance/Angle/Trim/Method/<Select first line>: *enter D*
Enter first chamfer distance <10.0000>: *enter* 20 ⏎
Enter second chamfer distance <20.0000>: *enter* 5 ⏎

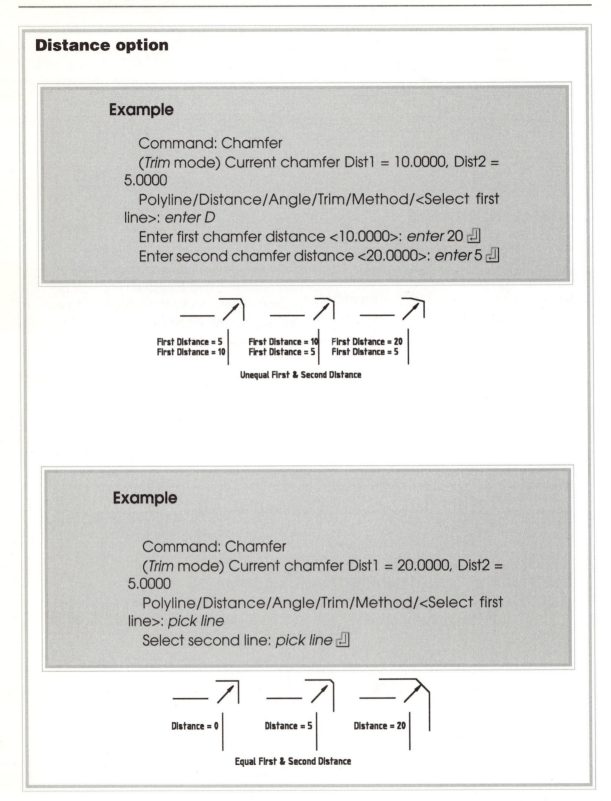

First Distance = 5   First Distance = 10   First Distance = 20
First Distance = 10   First Distance = 5   First Distance = 5

Unequal First & Second Distance

Example

Command: Chamfer
(*Trim* mode) Current chamfer Dist1 = 20.0000, Dist2 = 5.0000
Polyline/Distance/Angle/Trim/Method/<Select first line>: *pick line*
Select second line: *pick line* ⏎

Distance = 0   Distance = 5   Distance = 20

Equal First & Second Distance

Angle Option

Example

Command: _Chamfer
(*Trim* mode) Current chamfer Dist1 = 20.0000, Dist2 = 5.0000
Polyline/Distance/Angle/Trim/Method/ <Select first line>: *enter A* ⏎
Enter chamfer length on the first line<20.0000>: *select default* ⏎
Enter chamfer angle from the first line<0>: *enter 60* ⏎

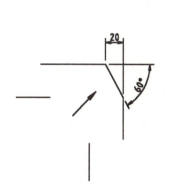

Trim Option

Example

Command: _Chamfer
(*Trim* mode) Current chamfer Dist1 = 5.0000, Dist2 = 5.0000
Polyline/Distance/Angle/Trim/Method/ <Select first line>: *enter T* ⏎
Trim/No trim<Trim>: *select default* ⏎
Polyline/Distance/Angle/Trim/Method/ <Select first line>: *pick line* ⏎
Select second line: *pick line* ⏎

Command: Chamfer
(*Trim* mode) Current chamfer Dist1 = 5.0000, Dist2 = 5.0000
Polyline/Distance/Angle/Trim/Method/ <Select first line>: *enter T* ⏎
Trim/No trim <Trim>: *enter N* ⏎

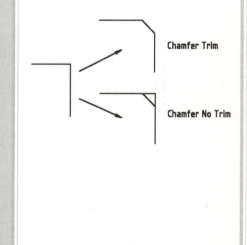

Method Option

Allows you to select either the *Distance* or *Angle* option.

Array command

Modify toolbar	⊞
Modify menu	**Array**
Command line entry	**Array**
Alias	**AR**

You can create copies of objects in a rectangular or polar (circular) pattern called an array. For rectangular arrays, you control the number of rows and columns and the distance between each. For polar arrays, you control the number of copies of the object and whether the copies are rotated. To create many regularly spaced objects arraying is faster than copying.

The *Array* dialogue box is presented when the Array command is activated.

Create using a closed polyline the drawing illustrated below. Use Grid and Snap set to 10. Do not include the dimensions.

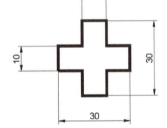

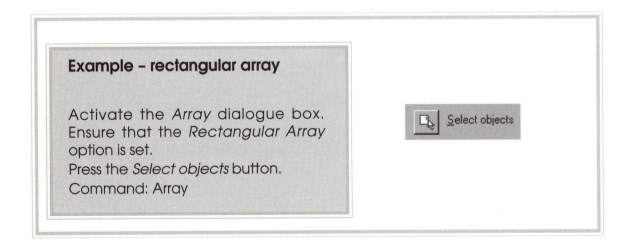

Example – rectangular array

Activate the *Array* dialogue box.
Ensure that the *Rectangular Array* option is set.
Press the *Select objects* button.
Command: Array

Select the cross drawing then right-click to accept your selection.

In the *Rows*: and *Columns*: edit boxes enter 4.

Enter 40 in the *Row offset* and *Column offset* edit boxes.

Press the *Preview* button to check your results, then *Accept*.

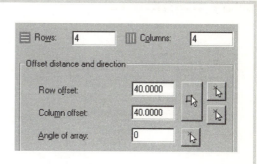

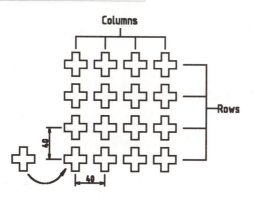

Type U at the command line to undo the Array command.

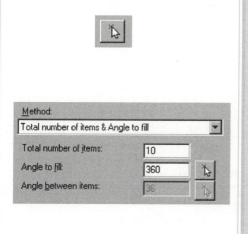

Example – polar array

Activate the Array dialogue box. Ensure that the Polar Array option is set.

Press the Select objects button and pick the cross.

Press the Centre point button and select a point on the drawing about which the cross will be arrayed.

Use the illustration as a guide.

Ensure that the Method box is set to:

Total number of items & Angle to fill.

Set Total number of items: to 10 and ensure Angle to *fill*: is set to 360.

Press the Preview button then Accept.

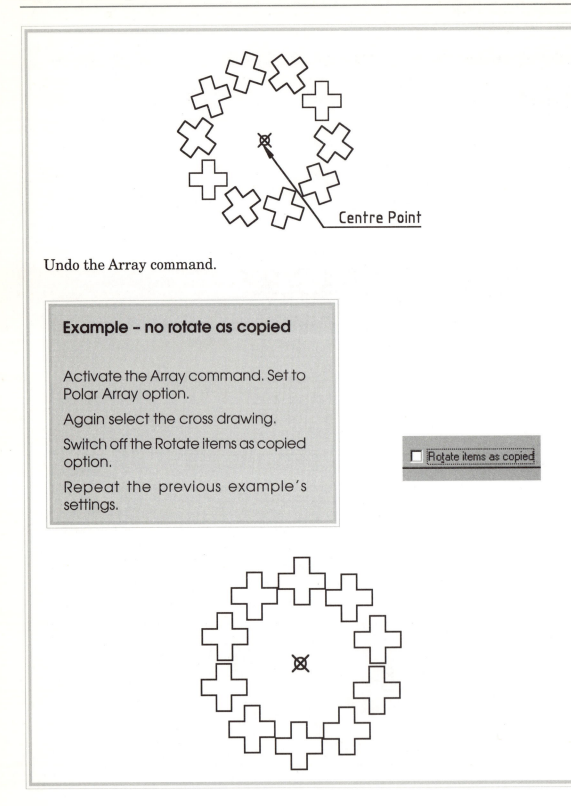

Centre Point

Undo the Array command.

Example – no rotate as copied

Activate the Array command. Set to Polar Array option.

Again select the cross drawing.

Switch off the Rotate items as copied option.

Repeat the previous example's settings.

☐ Rotate items as copied

Exercises using the Array command

Attempt the following Array exercises.

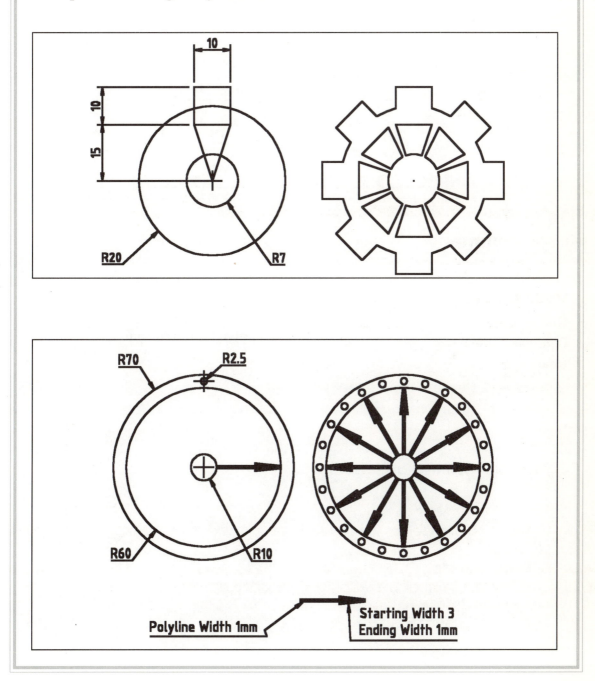

Use the Array and Mirror commands to complete the following exercise

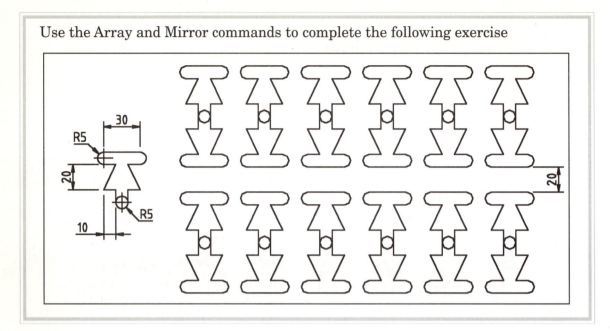

Mirror command

Modify toolbar	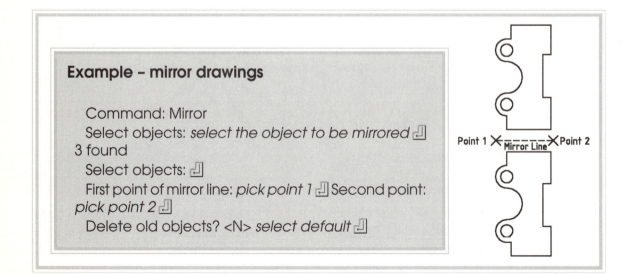
Modify menu	**Mirror**
Command line entry	**Mirror**
Alias	**MI**

Lets you make mirror images of existing objects about a mirror line.

Example – mirror drawings

Command: Mirror
Select objects: *select the object to be mirrored* ↵
3 found
Select objects: ↵
First point of mirror line: *pick point 1* ↵ Second point:
pick point 2 ↵
Delete old objects? <N> *select default* ↵

The mirror line can be at any angle and length. The user has the option to delete the originally selected objects if only the mirrored version is required.

The Mirrtext variable

The Mirror command will create a mirror image of alpha-numerical characters if so desired, but mirrored text may not always be the required outcome. The *Mirrtext* variable controls whether alpha-numerical text is mirrored.

Entering *Mirrtext* at the command line will result in the following prompt:

Enter new value for *Mirrtext* <0>:

Mirrtext set to 1 will result in alpha-numerical characters being mirrored.	Alpha-Numeric 12345 ABCDE	ɔiɿⱻmuИ-ɒʜqlA ƎᗡƆᙠA ટ4Ɛ21
Mirrtext set to 0 will result in alpha-numerical characters not mirrored.	Alpha-Numeric 12345 ABCDE	Alpha-Numeric 12345 ABCDE

Exercises – using the Mirror command

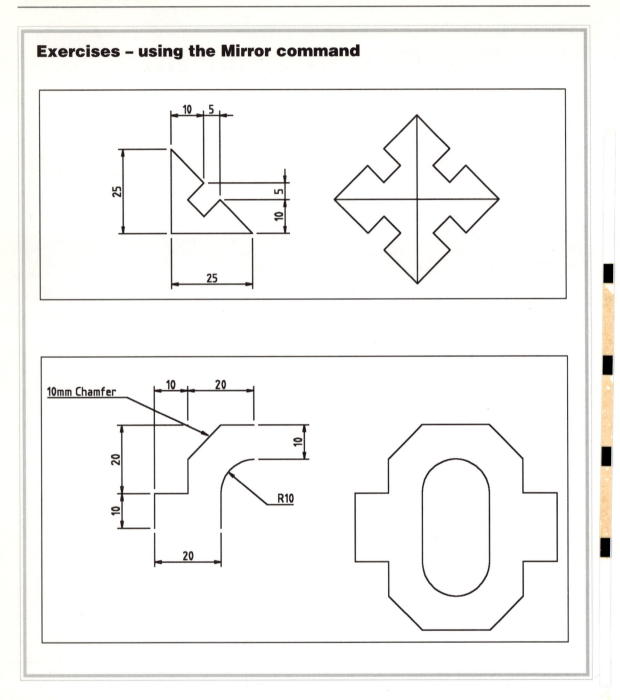

Practical assignment 2

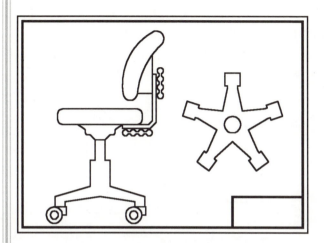

The following assignment will build upon the work undertaken within Practical assignment 1 by developing the swivel chair components, using a variety of modify commands to produce the drawing shown.

Open drawing file Swivel_chair.

1 Chair backrest

Zoom a window around the backrest objects.

Use the *Trim* command to produce the final drawing shape.

Move the backrest using the base point indicated to location 135,135.

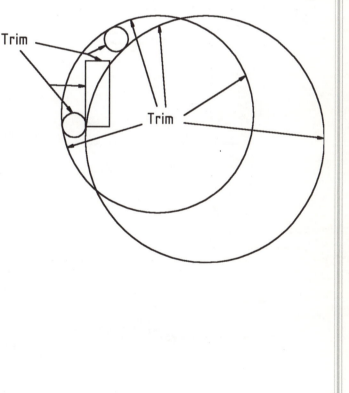

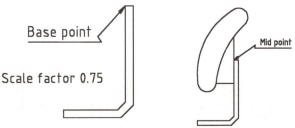

Chamfer distance 5mm

Rotation point (90°)

Chamfer distance 8mm

2.   Backrest bracket

Zoom a window around the backrest bracket.

Chamfer the bracket corners to the dimensions shown.

Rotate the bracket 90° and *Scale* to a factor of 0.75.

Using the base point indicated, *Move* the bracket to the mid-point at the rear of the backrest.

Base point

Scale factor 0.75

Mid point

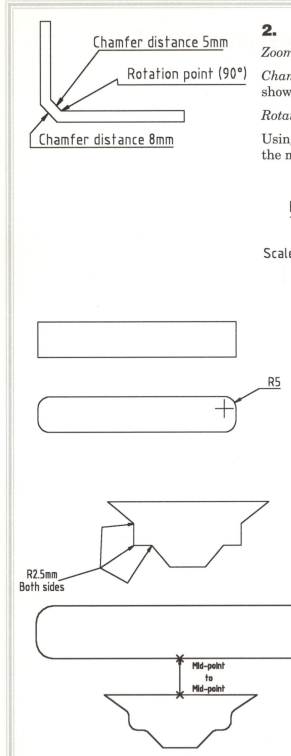

R5

R2.5mm
Both sides

Mid-point
to
Mid-point

3. Chair seat

Zoom a window around the chair seat.

Fillet each of the corner 5mm radius.

4. Swivel bracket

Zoom a window around the swivel bracket.

Fillet the corners indicated to a radius of 2.5mm

Move the swivel bracket from the top mid-point to the mid-point of the base of the seat.

Move both the swivel bracket and seat from the base point indicated to location 105,105.

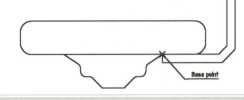

Base point

5. Chair base

Zoom a window around the chair base.

Stretch the base leg 5mm to the right.

Mirror the polyline about the end-points shown.

Move the base to the mid-point of the swivel bracket as shown.

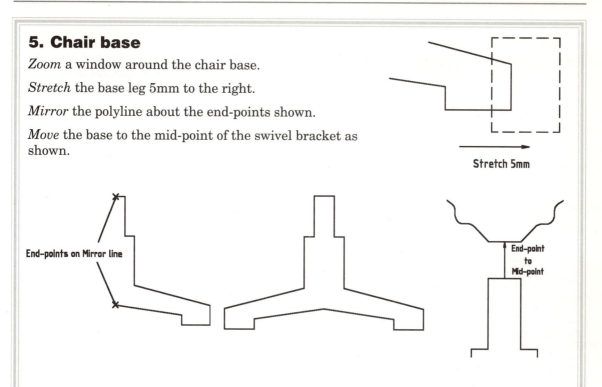

Stretch 5mm

End-points on Mirror line

End-point
to
Mid-point

6. Wheel castor

Zoom around the castor.

Trim the portions of the circle and rectangle shown.

Fillet the two remaining rectangle corners 2.5mm radius.

Move the castor base point mid-point to mid-point with the base leg.

Copy the castor to the opposite leg.

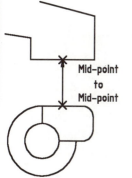

Mid-point
to
Mid-point

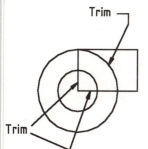

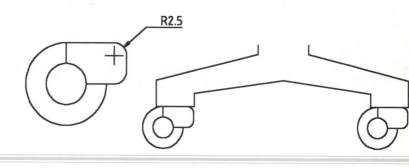

Trim

Trim

R2.5

87

7. Locking handle

Zoom around the handle objects.

Trim the portions of the circles indicated.

Move the handle to location 120,100 using the base point indicated.

Copy, Rotate 90° and Move a second handle to location 140,145.

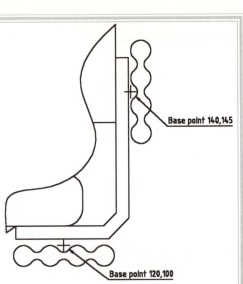

Base point 140,145

Trim •

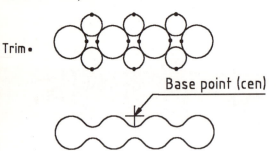

Base point (cen)

Base point 120,100

8. Base plan

Zoom around the base plan.

Use the centre of the circle as a base point and *Move* the plan objects to location 215,105.

Rotate the objects –90°.

Polar *Array* the base objects around the circle centre five times.

Trim the portions of the objects that overlap.

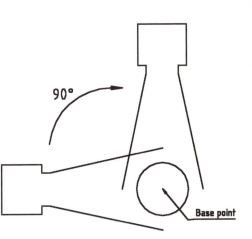

90°

Base point

Your drawing should look like the following illustration:

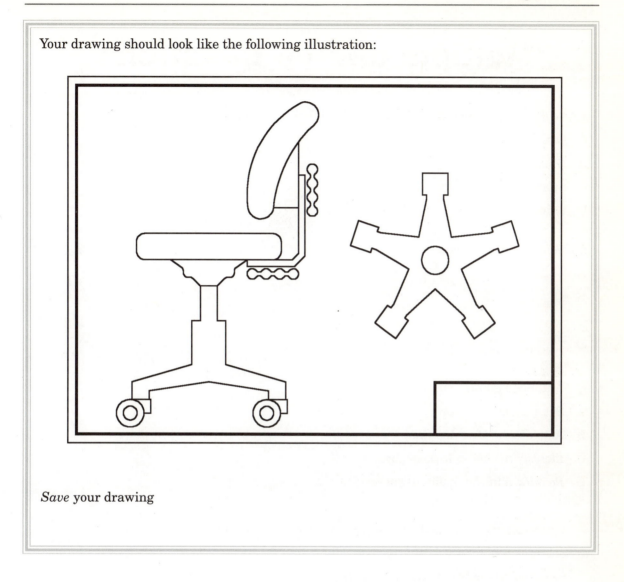

Save your drawing

Multiple choice questions

1 Which mode cannot be activated from the Drawing Aids dialogue box?

A Ortho

B Object Snap

C Snap

D Grid

2 Which of the following is a Rectangular Array?

A

B

C

D

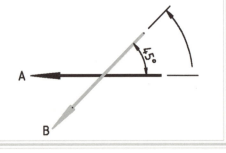

3 AutoSnap allows you to:

A Set a circle diameter

B Activate the Grid and Snap modes

C Identify the object to be snapped to

D Position a line endpoint in space

4 When selecting objects the Crossing box allows you to only:

A Select objects totally within the box

B Select objects only outside the box

C Select objects only touched by the box

D Select objects touched and within the box

5 At what angle has the arrow been rotated?

A −45°

B 45°

C −135°

D 135°

6   Which command would most efficiently create and position the circles shown?

A   Circle, Copy & Rotate option

B   Polar Array

C   Rectangular Array

D   Circle, Copy & Move option

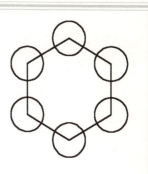

7   Which statement is true?

A   The Offset command can only be applied to closed objects

B   The Offset command only works on circles

C   The Offset command will only apply to the first selected object

D   The Offset command alias is OFF

8   In order to use the Stretch option you must select objects to be stretched using:

A   Picked points using the Fence selection option

B   A Crossing box

C   The screen cursor to select all objects individually

D   A Window

Chapter 5

Manipulating the drawing view

One of the most important features of a CAD system is to be able to manipulate the current view of the drawing. As a drawing develops, you will need to zoom in and out to view magnified parts of the drawing and to pan around to locate a particular area quickly. The Aerial view feature is a fast navigation tool that allows you to display specific areas quickly.

Objectives

At the end of this chapter you will be able to:

▷   Apply various Zoom options.

▷   Pan around a drawing.

▷   Use the Aerial View feature.

New commands

▷	Zoom	Z		▷	Properties	CH
▷	Pan	P		▷	Dtext	DT
▷	Aerial view	AV		▷	Mtext	MT
▷	Editing text	ED		▷	Spell Check	SP

Zoom command

Standard toolbar	
View menu	**Zoom**
Command line entry	**Zoom**
Alias	**Z**

The Zoom command allows you to control the viewing area of the graphics screen. The Zoom command can be activated from the *View* menu. The Zoom tools located on the Standard toolbar *include Zoom Realtime, Zoom Window* and *Zoom Previous*.

Selection of a tool with a triangle at its lower right corner will activate a drop down toolbar containing alternative zoom options. These include:

Zoom real-time

Hold down the pick button and move vertically upwards to zoom in 100%. Conversely, moving down to the bottom of the window zooms out by 100%.

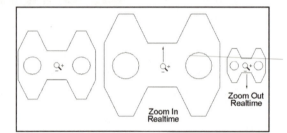

Zoom window

Displays the area specified by two opposite corners of a rectangle.

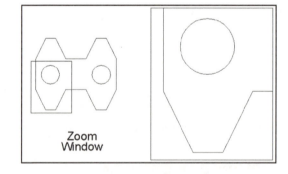

Zoom dynamic

Zooms to display the generated portion of the drawing with a view box. Generates a view box that represents your viewport. You can shrink or enlarge and move the view box around the drawing.

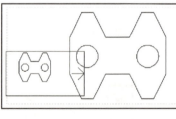
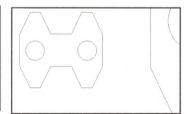

Zoom scale

The value you enter is relative to the limits of the drawing. For example, entering 2 doubles the apparent display size of any objects from what it would be if you were zoomed to the limits of the drawing.

If you enter a value followed by x, AutoCAD specifies the scale relative to the current view. For example, entering .5x causes each object to be displayed at half its current size on the screen.

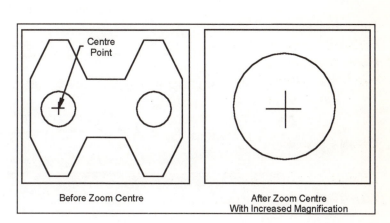

Zoom centre

Zooms to display a window defined by a centre point and a magnification value or height. A smaller value for the height increases the magnification. A larger value decreases the magnification.

Zoom in

Magnifies the display by a factor of 2.

Zoom out

Magnifies the display by a factor of 0.5.

Zoom all

Zooms to the drawing limits or current extents, whichever is greater.

Zoom extents

Zooms to display the drawing extents.

Pan command

Standard toolbar	
View menu	**Pan**
Command line entry	**Pan**
Alias	**P**

The Pan command is used to displace the drawing display. When activated, the cursor changes to a hand cursor. By holding down the pick button on the pointing device the drawing display is moved in the same direction as the cursor.

Example – using the zoom and pan commands

Open drawing file *db_samp.dwg* from the *Sample* folder within the *AutoCAD 2000 directory*.

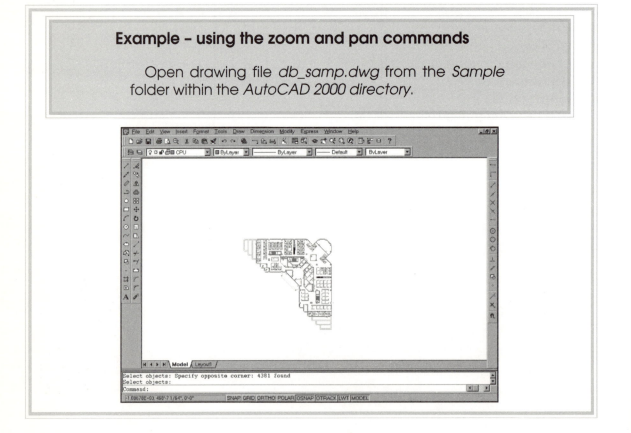

The drawing requires a zoom in order to fill the graphics area. Activate the *Zoom Window* tool in the *Standard* tool bar and place a window around the drawing.

Use the Real-time *Zoom and Pan* to view the central area of the drawing.

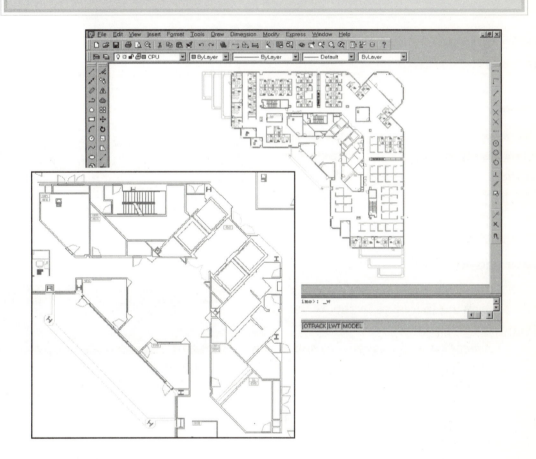

When you are satisfied with the required view, right click your mouse to activate the following menu.

Exit

✓ Pan
Zoom
3D Orbit

Zoom Window
Zoom Original
Zoom Extents

Select *Exit* to end *Zoom and Pan.*
Experiment with the other Zoom options to better understand their functions. When finished *Zoom Extents* to make the drawing fill the screen.

Aerial view

Standard toolbar	
View menu	**Aerial view**
Command line entry	**DSviewer**
Alias	**AV**

Aerial View is a navigation tool that allows you to display a view of the drawing in a separate window so that you can quickly move to that area. The Aerial View window can be kept active as you work enabling you to zoom and pan without choosing a menu or tool bar option.

View menu

Zoom In – The magnification of the drawing increases in the Aerial View by a factor of 2.

Zoom Out – The magnification of the drawing decreases in the Aerial View by a factor of 2.

Global – Displays the entire drawing and the current view in the Aerial View window.

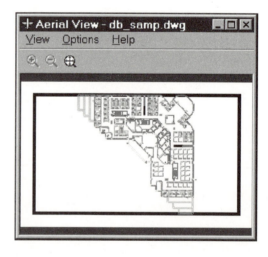

Options menu

Auto Viewport – When active, the model space view of the active viewport is displayed automatically.

Dynamic Update – When active, the Aerial View window updates while you edit the drawing.

Real-time Zoom – The drawing area updates in real time when you zoom using the Aerial View window.

Example – panning and zooming using aerial view

1.   Activate the Aerial View window.

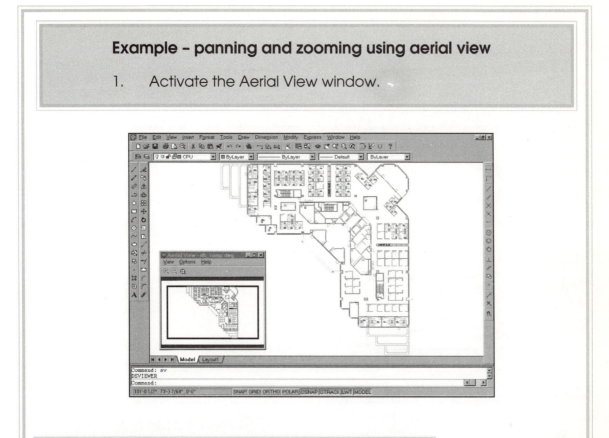

2.   Left click your cursor in the centre of the aerial view window. A rectangular viewing box will appear. Notice as you move this viewing box, the drawing is panning about the screen.

3.   Left click your cursor again and the X is replaced by an arrow. Move the cursor whilst holding down the left mouse button and the rectangular viewing box will increase and decrease in size. The rectangle represents the screen area, therefore the smaller the rectangular box, the greater the screen magnification.

4.   Reduce the rectangle to approximately half its current size and left click your mouse again.

5.   Pan the viewing box to the lower right of the drawing then right click your mouse to end the process.

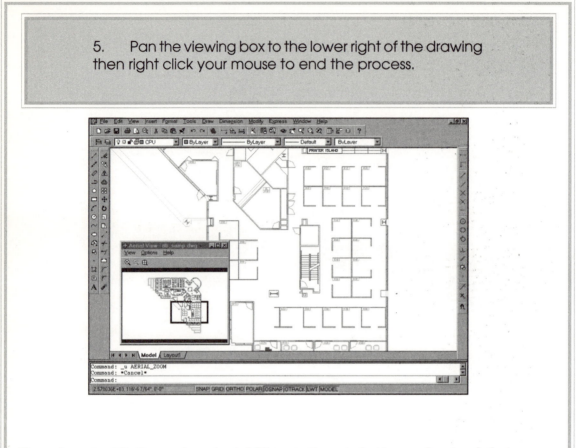

Experiment with the various Aerial View options to better understand their particular functions.

Chapter 6

More draw commands

The following chapter introduces more drawing commands that will further develop your CAD skills.

Objectives

At the end of this chapter you will be able to:

▷   Be familiar with the commands shown below.

New commands

▷	Ray			▷	Sketch		
▷	Construction Line	Xline		▷	Ellipse	EL	
▷	Multiline	Mline		▷	Donut	DO	
▷	Arc	A		▷	Point	PO	
▷	Spline	SPL					

Ray command

Draw menu	**Ray**
Command line entry	**Ray**

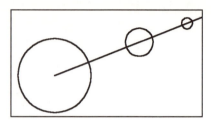

A Ray creates a construction line from a starting point which extends to infinity. A Ray has no effect on the extents of a drawing but can be moved, rotated and copied.

Construction line

Draw toolbar	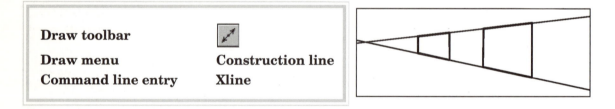
Draw menu	**Construction line**
Command line entry	**Xline**

A Construction line (or Xline) is specified by two points and extends into infinity in both directions. A construction line has no effect on the extents of a drawing but can be moved, rotated and copied.

Multiline command

Draw toolbar	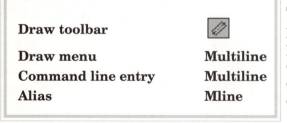
Draw menu	**Multiline**
Command line entry	**Multiline**
Alias	**Mline**

The Multiline command creates multiple parallel lines called elements. Elements are positioned by offsetting each one from the origin of the multiline. You can set the colour and linetype as well as the offset distance between each element.

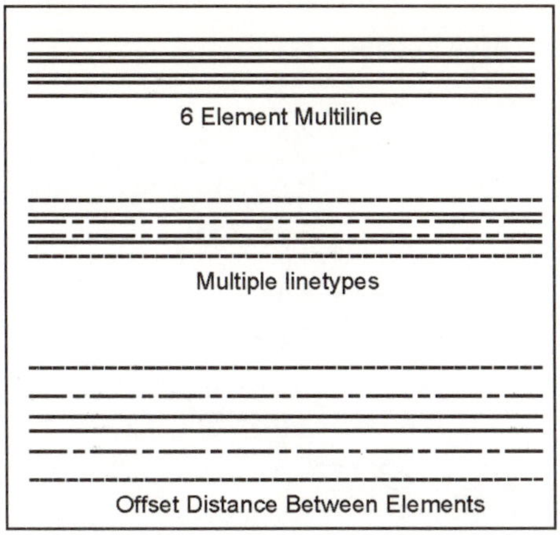

6 Element Multiline

Multiple linetypes

Offset Distance Between Elements

- Multilines can contain between 2 and 16 elements.
- Each element can be set to a specific colour and linetype.
- The offset distance between elements can be varied.

The *Multiline Styles* dialogue box is used to create multiline styles.

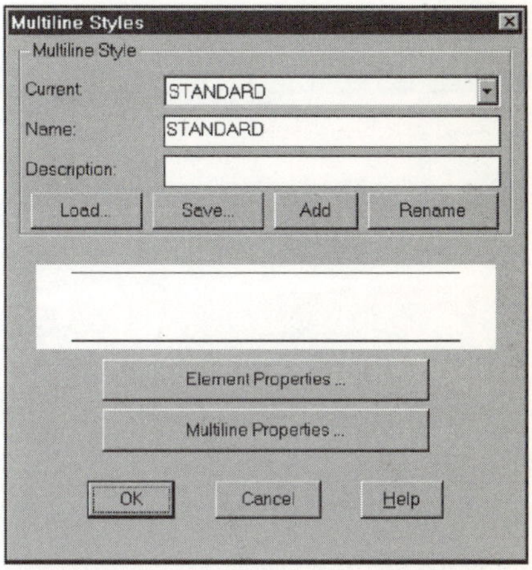

Element Properties is used to define each element's offset distance, colour and linetype.

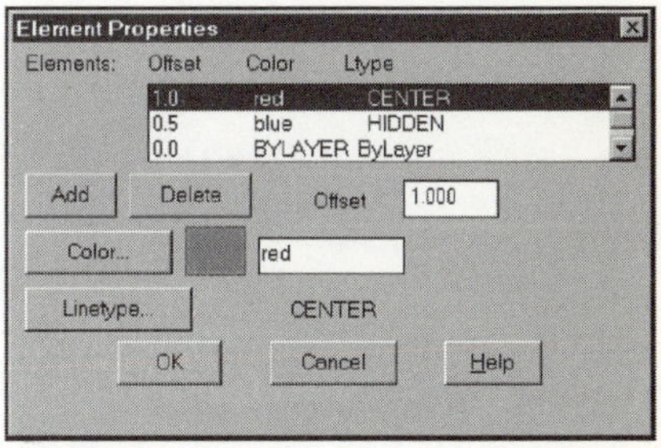

If required, multilines can include joints which appear at each vertex and caps at either or both ends of the line.

The *Multiline Properties* dialogue box controls the appearance of Joints, Caps and the Fill option.

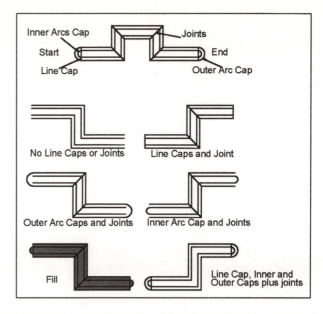

Example

1.   Start a *New* drawing using the *Start from Scratch – Metric* option.

2.   From the *Draw* menu select *Multiline*. Enter the following:

Command: mline

Current settings: Justification = Top, Scale = 20.00, Style = *Standard*

Specify start point or (Justification/Scale/STyle): *enter* 50,50

Specify next point: *enter* @100<0

Specify next point or (Undo): *enter* @100<90

Specify next point or (Close/Undo): *enter* @50<180

Specify next point or (Close/Undo): *enter* C (Close)

Example – creating a new multiline style

1   Select *Multiline Style...* from the *Format* menu.

2   Enter *Mltest1* in the style *Name* box then press the *Add* button.

3.   Press the *Element Properties...* button to activate the *Element Properties* dialogue box. There are currently only two elements specified. You will add a further two elements.

4.   Press the *Add* button twice. Two new elements will appear.

5.   Whilst one of the elements is selected, enter a value of 1 in the *Offset* edit box then press *Return*.

6.   Select and highlight the second element with the offset value of 0. Enter an *Offset* value of –1 (minus 1). The element list will now show:

1.0	Bylayer	Bylayer
0.5	Bylayer	Bylayer
–0.5	Bylayer	Bylayer
–1.0	Bylayer	Bylayer

7.   Select element *1.0* then press the colour swatch and pick colour *Blue*.

8.   Select element *–1.0* press the colour swatch and again pick the colour *Blue*.

9.   Press *OK* to exit. The graphics illustration will indicate the new multiline style.

More draw commands

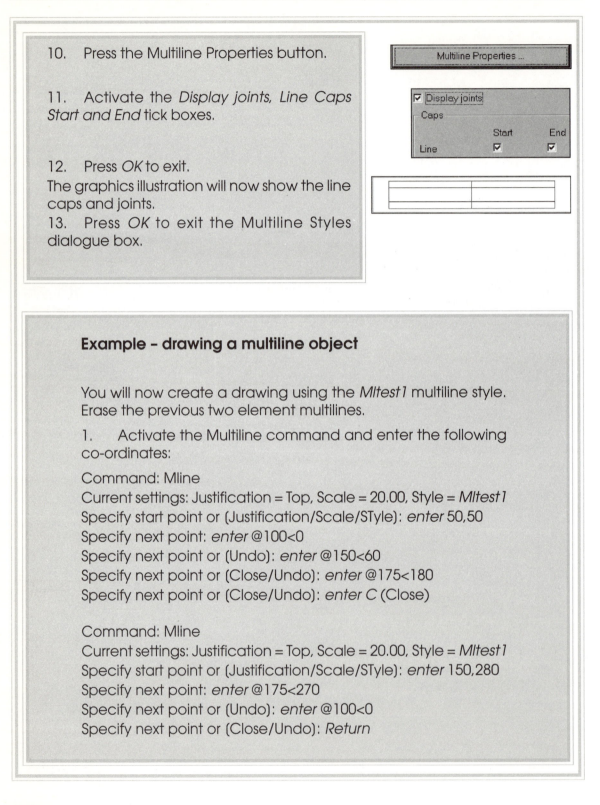

10. Press the Multiline Properties button.

11. Activate the *Display joints, Line Caps Start and End* tick boxes.

12. Press *OK* to exit.
The graphics illustration will now show the line caps and joints.

13. Press *OK* to exit the Multiline Styles dialogue box.

Example – drawing a multiline object

You will now create a drawing using the *Mltest1* multiline style. Erase the previous two element multilines.

1. Activate the Multiline command and enter the following co-ordinates:

Command: Mline
Current settings: Justification = Top, Scale = 20.00, Style = *Mltest1*
Specify start point or (Justification/Scale/STyle): *enter* 50,50
Specify next point: *enter* @100<0
Specify next point or (Undo): *enter* @150<60
Specify next point or (Close/Undo): *enter* @175<180
Specify next point or (Close/Undo): *enter* C (Close)

Command: Mline
Current settings: Justification = Top, Scale = 20.00, Style = *Mltest1*
Specify start point or (Justification/Scale/STyle): *enter* 150,280
Specify next point: *enter* @175<270
Specify next point or (Undo): *enter* @100<0
Specify next point or (Close/Undo): *Return*

Modifying multilines

Modify II toolbar	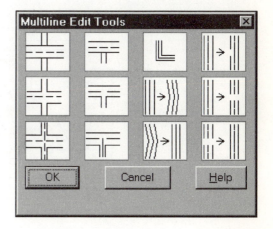
Modify menu	**Multiline**
Command line entry	**MLEDIT**

The *Multiline Edit Tools* dialogue box is used to create and modify multiline patterns.

Each column illustrates a particular editing process. These include multilines that cross, multilines that form a tee, corner joints and vertices, and multilines to be cut or joined.

Example

1.   From the *Modify* menu select *Multiline…* to activate the *Multiline Edit Tools* dialogue box.
2.   Select the *Merged Cross* icon.

3.   Select Multiline 1 and 2 as indicated.

4.   Activate the *Multiline Edit Tools* dialogue box again and select the *Merged Tee* icon.

5.   Select the multilines 1 and 2 as indicated.

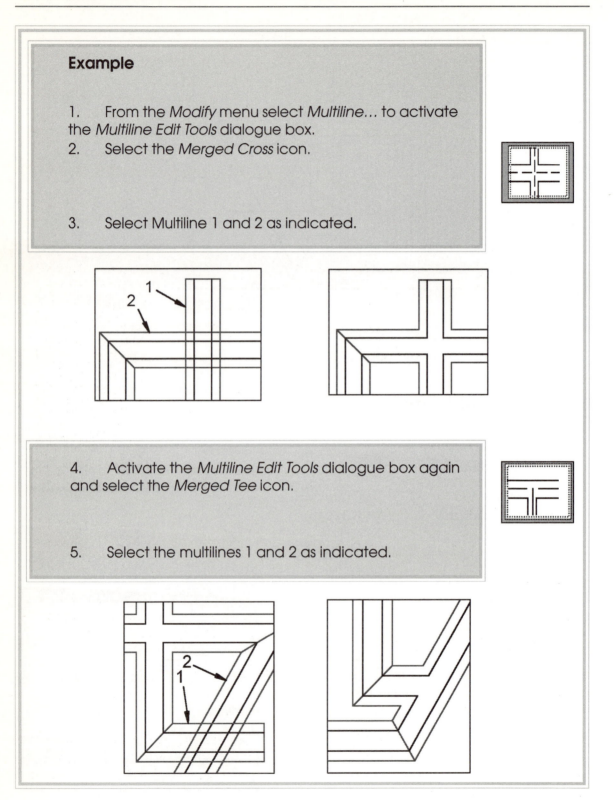

Arc command

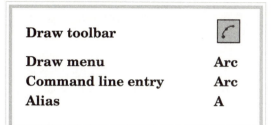

Draw toolbar	
Draw menu	Arc
Command line entry	Arc
Alias	A

There are a variety of options available to create an Arc. The most basic way to define an arc is to use the 3 point option.

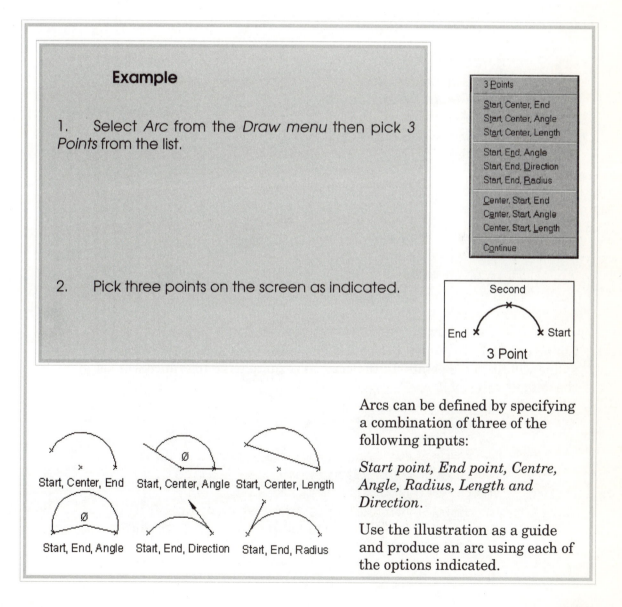

Example

1. Select *Arc* from the *Draw menu* then pick *3 Points* from the list.

2. Pick three points on the screen as indicated.

Arcs can be defined by specifying a combination of three of the following inputs:

Start point, End point, Centre, Angle, Radius, Length and Direction.

Use the illustration as a guide and produce an arc using each of the options indicated.

Spline command

Draw toolbar	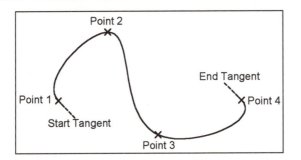
Draw menu	Spline
Command line entry	Spline
Alias	SPL

A Spline is a quadratic or *NURBS* curve. The Spline command fits a smooth curve through a set of points. Splines are useful for drawing contour lines for mapping purposes or where smooth surfaces are required.

Example

1. Select *Spline* from the *Draw* menu.
2. Use the cursor to specify points 1–4 as indicated then press *Return*.
You will be prompted to specify the start and end tangents.
3. Move the cursor in the directions indicated in the illustration.

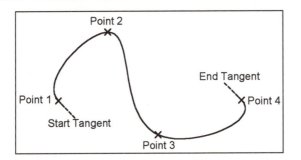

Spline curves have a tolerance value. The lower the tolerance the closer the spline will fit to the points. In the previous example the tolerance was set at the default value of 0.

Example – spline tolerance

1. Repeat the Spline command as above. After you have specified the second point, the Tolerance option will appear on the command line.
2. Enter *T* for Tolerance and set a value of 10. Complete the Spline, including the start and end tangents as before.

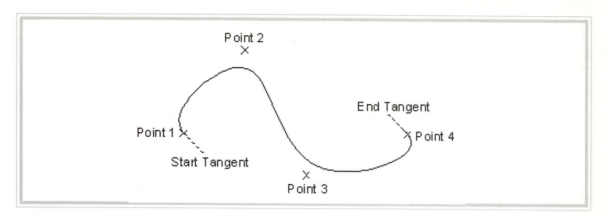

Sketch command

Command line entry	Sketch

A Sketch creates a series of line segments.

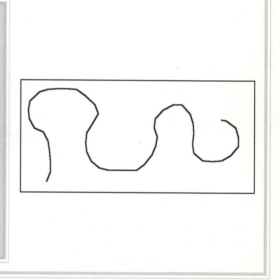

Example

1. At the command line enter *Sketch*.
2. The *Record Increment* prompt determines the length of each of the line segments that make up the sketch. Enter a value of 1.
3. At the <Pen down> prompt left click and sketch the shape required. Left click again for <Pen up> to end sequence. Press *Return* to end.

Ellipse command

Draw toolbar	⬭
Draw menu	**Ellipse**
Command line entry	**Ellipse**
Alias	**EL**

An Ellipse can be created by using one of the following options:

Center, Axis Endpoint, Arc or *Isocircle.*

Center
Axis, End

Arc

Example - center

Command: _ellipse
Specify axis endpoint of ellipse or (Arc/Center): *enter C*
Specify center of ellipse: <Snap on> *pick point 1*
Specify endpoint of axis: *pick point 2*
Specify distance to other axis or (Rotation): *pick point 3*

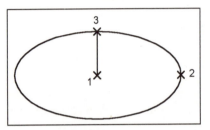

Example - axis endpoint

Command: _ellipse
Specify axis endpoint of ellipse or (Arc/Center): *pick point 1*
Specify other endpoint of axis: *pick point 2*
Specify distance to other axis or (Rotation): *pick point 3*

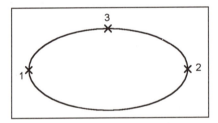

Example - arc

Command: _ellipse
Specify axis endpoint of ellipse or (Arc/Center): *enter A (Arc)*
Specify axis endpoint of elliptical arc or (Center): *pick point 1*
Specify other endpoint of axis: *pick point 2*
Specify distance to other axis or (Rotation): *pick point 3*
Specify start angle or (Parameter): *enter 330*
Specify end angle or (Parameter/Included angle): *enter 210*

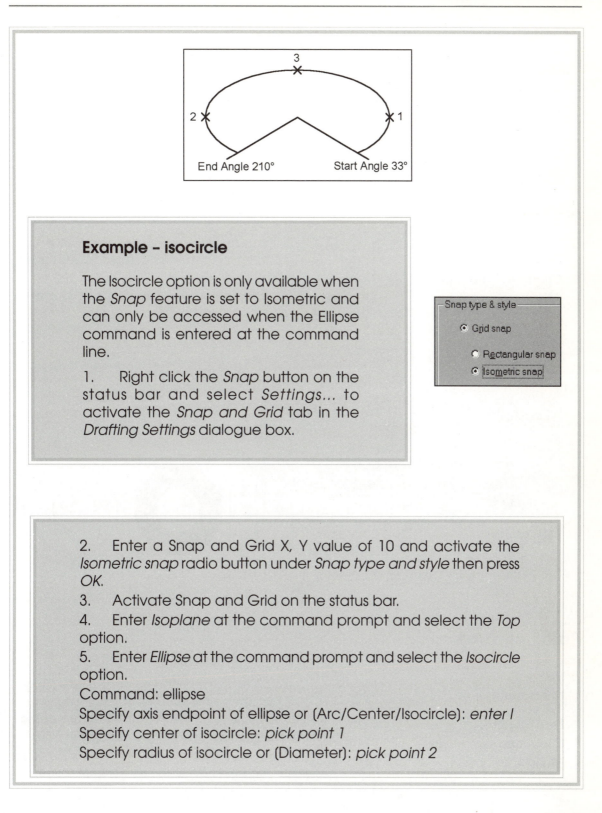

Example – isocircle

The Isocircle option is only available when the *Snap* feature is set to Isometric and can only be accessed when the Ellipse command is entered at the command line.

1.   Right click the *Snap* button on the status bar and select *Settings...* to activate the *Snap and Grid* tab in the *Drafting Settings* dialogue box.

2.   Enter a Snap and Grid X, Y value of 10 and activate the *Isometric snap* radio button under *Snap type and style* then press *OK*.

3.   Activate Snap and Grid on the status bar.

4.   Enter *Isoplane* at the command prompt and select the *Top* option.

5.   Enter *Ellipse* at the command prompt and select the *Isocircle* option.

Command: ellipse

Specify axis endpoint of ellipse or (Arc/Center/Isocircle): *enter I*

Specify center of isocircle: *pick point 1*

Specify radius of isocircle or (Diameter): *pick point 2*

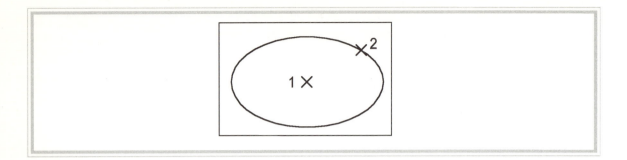

Donut command

Draw menu	**Donut**
Command line entry	**Donut**
Alias	**DO**

A donut is a circular closed polyline. A donut specification is defined by an inside and outside diameter. The donut can be filled to appear solid or clear depending on the setting of the Fill command.

Example

1. Activate the *Donut* command.
2. Enter an inside diameter of 10 and an outside diameter of 20.
3. At the command prompt enter *Fill* and set to *Off*.
4. Repeat the *Donut* command.

FILL On   FILL Off

Point command

Draw toolbar	▣
Draw menu	**Point**
Command line entry	**Point**
Alias	**PO**

Point objects are useful as node points that you can snap to and offset objects from. There are a variety of Point styles available. Point style and size are set using the *Point Style* dialogue box which is activated by selecting *Point Style* in the *Format* menu.

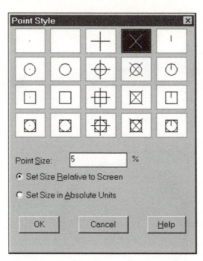

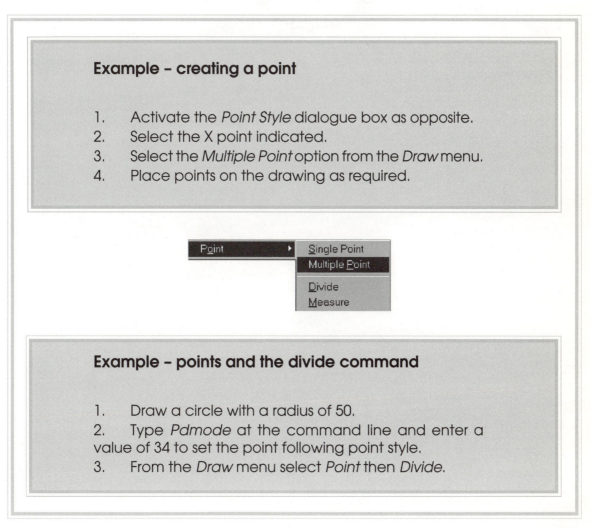

Example – creating a point

1. Activate the *Point Style* dialogue box as opposite.
2. Select the X point indicated.
3. Select the *Multiple Point* option from the *Draw* menu.
4. Place points on the drawing as required.

Example – points and the divide command

1. Draw a circle with a radius of 50.
2. Type *Pdmode* at the command line and enter a value of 34 to set the point following point style.
3. From the *Draw* menu select *Point* then *Divide*.

4.   Command: _divide. Select object to divide: *pick the circle*. Enter the number of segments or (Block): *enter* 20.

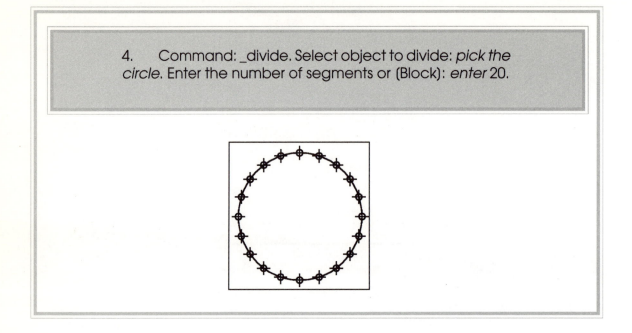

Chapter 7

More modify commands

The following chapter introduces more editing commands that will further develop your CAD skills.

Objectives

At the end of this chapter you will be able to:

▷ Be familiar with the commands shown below.

New commands

▷   Polyline edit   PE

▷   Scale   SC

▷   Lengthen   LEN

▷   Break   BR

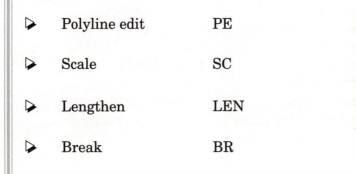

Modifying polylines

Modify toolbar	
Modify menu	**Polyline**
Command line entry	**PEDIT**
Alias	**PE**

Polylines can be edited using the *Pedit* command. The vertices within a polyline can be moved, added and deleted. Polyline width can be changed along the entire polyline length or individually between segments. The polyline linetype style can be altered and you can create a linear approximation of a spline from a polyline.

Example – modifying polyline vertices

1.   Start a *New* drawing. Select the *Use a Template* option and pick *Student.dwt*.
2.   Create the drawing shown using the *Polyline* command. Draw the shape in a clockwise direction.

3.   From the *Modify* menu select *Polyline* and select the polyline when prompted.
4.   Enter *E* to select *Edit Vertex* from the list of options present on the command line.
5.   A cross will appear in the lower left corner. Press the *Space bar* once to move the cross to the next vertex position (top left corner).

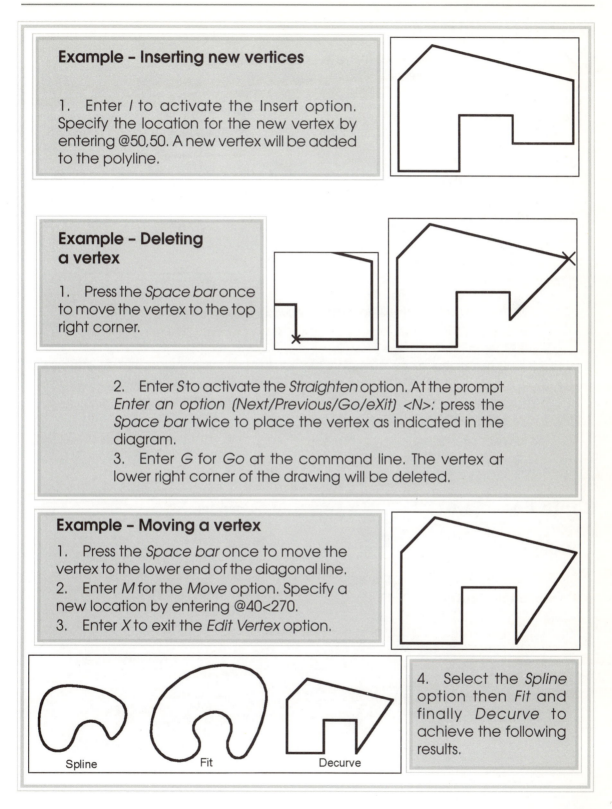

Example – Inserting new vertices

1. Enter *I* to activate the Insert option. Specify the location for the new vertex by entering @50,50. A new vertex will be added to the polyline.

Example – Deleting a vertex

1. Press the *Space bar* once to move the vertex to the top right corner.

2. Enter *S* to activate the *Straighten* option. At the prompt *Enter an option (Next/Previous/Go/eXit) <N>:* press the *Space bar* twice to place the vertex as indicated in the diagram.
3. Enter *G* for *Go* at the command line. The vertex at lower right corner of the drawing will be deleted.

Example – Moving a vertex

1. Press the *Space bar* once to move the vertex to the lower end of the diagonal line.
2. Enter *M* for the *Move* option. Specify a new location by entering @40<270.
3. Enter *X* to exit the *Edit Vertex* option.

4. Select the *Spline* option then *Fit* and finally *Decurve* to achieve the following results.

Spline   Fit   Decurve

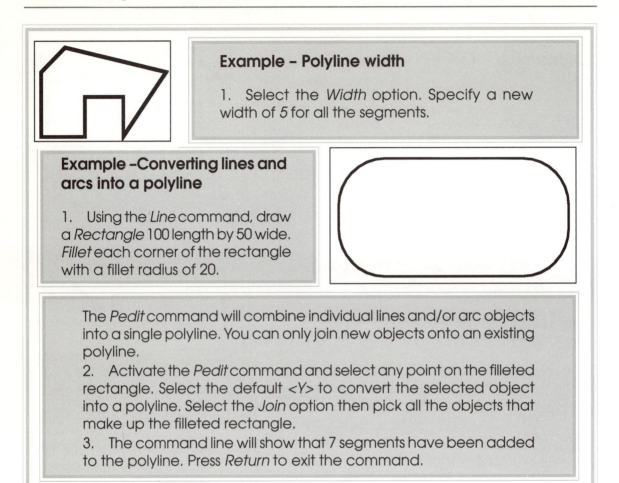

Example – Polyline width

1. Select the *Width* option. Specify a new width of *5* for all the segments.

Example –Converting lines and arcs into a polyline

1. Using the *Line* command, draw a *Rectangle* 100 length by 50 wide. *Fillet* each corner of the rectangle with a fillet radius of 20.

The *Pedit* command will combine individual lines and/or arc objects into a single polyline. You can only join new objects onto an existing polyline.

2. Activate the *Pedit* command and select any point on the filleted rectangle. Select the default *<Y>* to convert the selected object into a polyline. Select the *Join* option then pick all the objects that make up the filleted rectangle.

3. The command line will show that 7 segments have been added to the polyline. Press *Return* to exit the command.

Scale command

Modify toolbar	
Modify menu	**Scale**
Command line entry	**Scale**
Alias	**SC**

Selected objects are enlarged or reduced about a base point. Scaling occurs equally in the X, Y, and Z directions.

Example

1. Select objects to be scaled.

2. Specify a base point about which the objects will be scaled.

3. Enter a scaling factor.

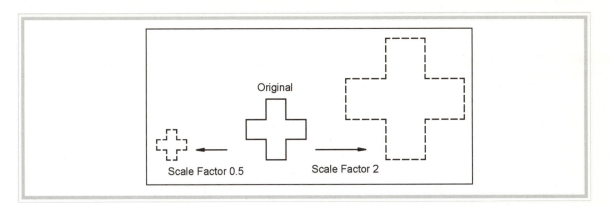

Original

Scale Factor 0.5

Scale Factor 2

Lengthen command

Modify toolbar	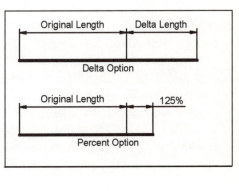
Modify menu	**Lengthen**
Command line entry	**Lengthen**
Alias	**LEN**

The *Lengthen* command is used to change the length of an object. There are four options: Delta, Percent, Total and Dynamic.

Example

Pick the object to be lengthened.

Delta option

Select the *Delta* option and enter the amount you want to lengthen the object.

Percent option

Enter the total percentage increase of the object where the original length equals 100%.

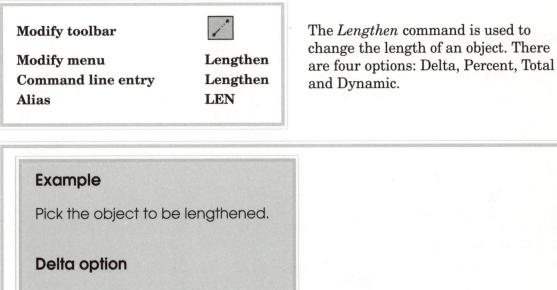

Original Length | Delta Length

Delta Option

Original Length | 125%

Percent Option

Total option

Specify the required total length of the original object.

Dynamic option

Changes the length of an object by dragging its endpoint to the required location.

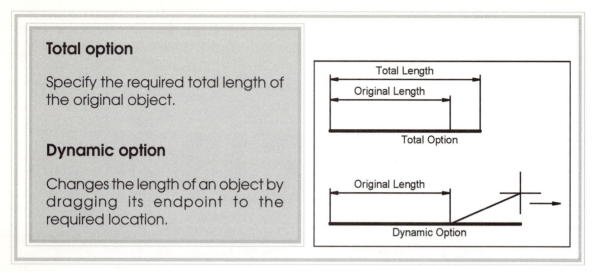

Break command

Modify toolbar	
Modify menu	**Break**
Command line entry	**Break**
Alias	**BR**

The *Break* command will erase a part of a selected object specified by two points. The *Break* command works in an anti-clockwise direction.

Example

1. Draw a *Circle* and *Rectangle* as indicated. The *Break* command will take the point on the object when it is selected, as the first break point. Entering *F* for First point, will allow you to select the two *Break* points on the object after it has been selected.

2. Select the *Circle*. Enter *F* for *First* point then select the approximate points indicated.

3. Repeat for the *Rectangle*.

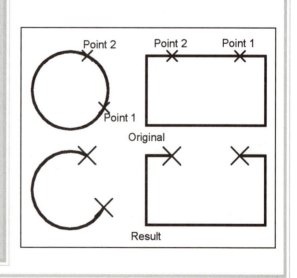

Chapter 8

Layers, linetypes and colours

Drawing can be controlled by the use of layers, linetypes and colours. A complex drawing may include text, construction lines, hidden detail, dimensions, various views, etc., all of which combined would present the reader with too much information. A drawing could easily become cluttered. By assigning aspects of your drawing to a particular layer, i.e., text to the text layer, you can make visible only that information you wish to present. This chapter will show you how to create new layers, allocating each layer its own specific colour and linetype. The layer options enable you to control the visibility, security and regeneration of these new layers.

Objectives

At the end of this chapter you will be able to:

▷   Create new layers.

▷   Allocate new layers, colour and linetype.

▷   Move objects between layers.

▷   Control layer visibility, security and regeneration.

New command

▷   Layer   LA

Layers

A complex drawing may contain a large number of elements including text, construction lines, hidden detail, dimensions, various views, etc. all of which could result in a cluttered and confusing presentation.

AutoCAD allows you to create as many layers as is necessary to organise your drawing.

Think of a layer as a sheet of overhead transparency onto which you can place specific objects such as text or dimensions. By switching these layers on or off you can determine the visibility of specific elements within the drawing.

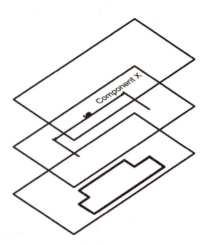

You can allocate each layer its own specific name, colour and linetype and specify whether objects on a layer can be edited. This enables you to plot different annotations of the same drawing by only displaying the layers that you require.

Controlling layers

Layers are created and controlled from the *Layer Properties Manager* Dialogue Box which can be located on the Object Properties toolbar

or by selecting *Layer...* from the *Format* pull down menu. This dialogue box is used to assign each layer a specific name, linetype and colour.

Alternatively you can access and alter a layer's state by using the Layer drop down menu located on the *Object Properties* toolbar.

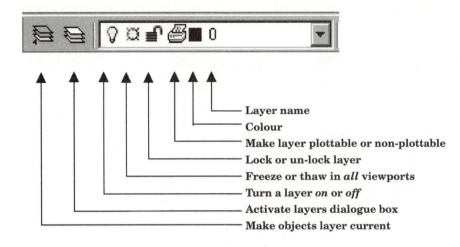

Layer name
Colour
Make layer plottable or non-plottable
Lock or un-lock layer
Freeze or thaw in *all* viewports
Turn a layer *on* or *off*
Activate layers dialogue box
Make objects layer current

You can use the *Layer* drop down menu to alter the state of existing layers in order to make a layer current, turn layers on and off, freeze and thaw layers, lock and unlock layers or make a layer plottable or non-plottable. You use this menu when you only want to perform an action on one layer at a time.

Layer properties

Layers have the following properties:

Layer Name – The name of layer, can be up to 31 characters in length, made up from characters, digits or special characters (dollar $, hyphen – and underscore_).

Visibility – Layers can turned either On or Off.

Freeze / Thaw – Controls the regeneration of the layer when a display is regenerated. Thawed layers are regenerated. Freezing unneeded layers increases regeneration speed.

Locked / Unlocked – Objects on locked layers are visible but cannot be edited. Locked layers can be current, can be drawn upon and linetype and colour can be changed.

Layer Plottable – Layers can be plottable or non-plottable. A non-plottable layer will be visible on the screen but will not appear in a hard plot.

Colour Number – The colour number determines the colour for visible layers. Default layer colour is 7 (white, though appears black on the screen!). Objects can override colours assigned to layers.

Linetype – Name of linetype defined for that layer. Several layers can use the same linetype. Objects can override the linetype assigned to a layer.

Lineweight – Displays the available lineweights that can be applied to a layer.

Example – set layer colour

Press the *Layers* button 🗇 to activate the *Layer Properties Manager* dialogue box.

Press the *New* button and enter the layer name *Construction*.

Name	On	Freeze...	L...	Color	Linetype	Lineweight	Plot Style	Plot
0	💡	☼	🔓	■ White	Continuous	—— Default	Color_7	🖨
Construction	💡	☼	🔓	■ Magenta	Continuous	—— Default	Color_6	🖨

Note – placing a comma(,) after a new layer name will force an additional new layer line to appear.

Select the colour name *White* on the Construction line.

This will activate the *Select Colour* dialogue box.

Use the cursor arrow to select the *Red* colour.

Elements constructed on or moved to the *Layer Construction* will now be *Red* in colour.

Example – set layer linetype

Select the Linetype Continuous on the *Construction* line.
This will activate the *Select Linetype* dialogue box.

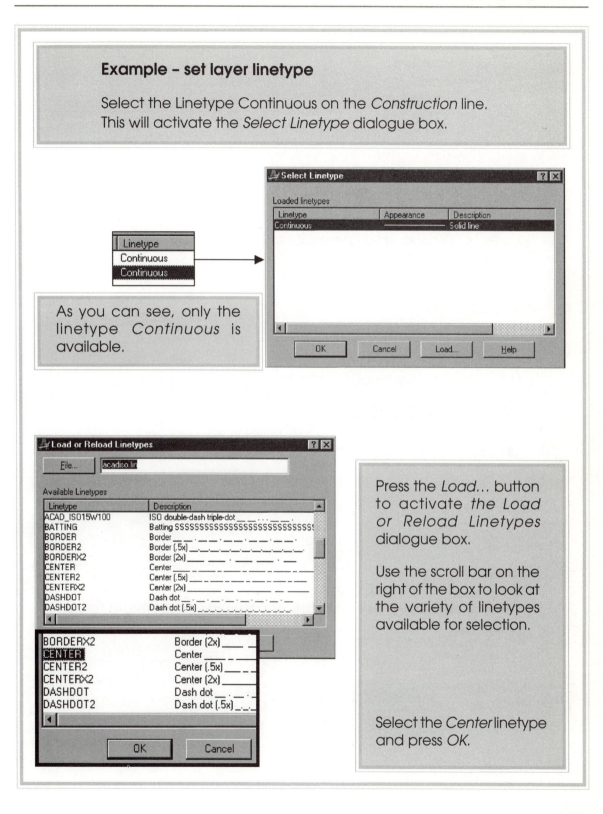

As you can see, only the linetype *Continuous* is available.

Press the *Load...* button to activate *the Load or Reload Linetypes* dialogue box.

Use the scroll bar on the right of the box to look at the variety of linetypes available for selection.

Select the *Center* linetype and press *OK*.

The *Center* linetype is now available for selection to be attached to layer *Construction*.

Pick the Linetype *Center*.

Loaded linetypes		
Linetype	Appearance	Description
CENTER	— · — · — · —	Center
Continuous	———————	Solid line

Example – set layer Lineweight

Select *Default* under Lineweight to activate the *Lineweight* dialogue box.

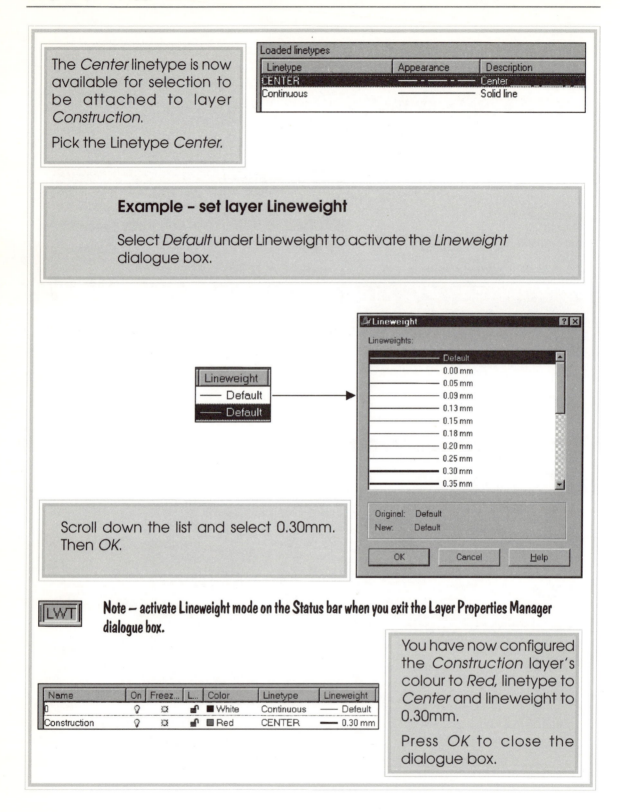

Scroll down the list and select 0.30mm. Then *OK*.

Note – *activate Lineweight mode on the Status bar when you exit the Layer Properties Manager dialogue box.*

You have now configured the *Construction* layer's colour to *Red*, linetype to *Center* and lineweight to 0.30mm.

Press *OK* to close the dialogue box.

Name	On	Freez...	L...	Color	Linetype	Lineweight
0	♀	☼	🔓	■ White	Continuous	— Default
Construction	♀	☼	🔓	■ Red	CENTER	— 0.30 mm

128

Exercise

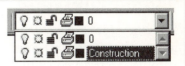

1. Click on to the Layer drop down menu and make *Construction* current.

2. Draw a rectangle on your screen and you will see that it's colour is *Red* and linetype set to *Center*.

3. Make Layer 0 current once again.

4. Draw another rectangle anywhere on your screen. It will default to white.

5. Freeze layer *Construction* using the layer pull down box, the red rectangle will disappear.

6. Thaw layer *Construction*, make it the current layer and Freeze layer 0.

Layer creation exercise

Using the table as a guide produce the following layers and assign the colours, linetypes and lineweights shown.

Layer name	Layer colour	Layer linetype	Lineweight
Circle	Red	Dashed	Default
Rectangle	Magenta	Hidden	0.30mm
Polygon	Blue	Border	0.50mm

Activate the *Circle* layer and draw the circles as indicated in the illustration. Repeat for the *Rectangle* and *Polygon* layers.

Use the Layer drop down menu to Freeze Layers *Rectangle* and *Polygon*.

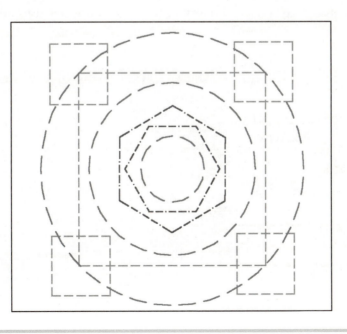

Loading linetypes

When you start a *New* drawing, the default linetype is set to *Bylayer*, that is, the linetype that is set to Layer 0. This linetype is continuous and is the only linetype available. You can see which linetypes are available by selecting the Linetype Control box in the *Object Properties* toolbar.

You can load alternative linetypes by selecting *Other...* to activate the *Linetype Manager* dialogue box, which can also be activated by selecting *Linetype...* from the *Format* pull down menu.

Press the *Load* button.

Drag down the scroll bar to see the variety of linetypes that are available for selection.

Example

Drag the scroll bar down and select the linetype *hidden*.
Press *OK*.

The *Hidden* linetype is now available for selection within the *Linetype Manager* dialogue box.

To make it active, select *Hidden* then press the *Current* button.

```
Linetype Manager                                              [?][X]
┌─ Linetype filters ──────────────────────┐   ┌─ Load... ─┐  ┌ Delete ┐
│ Show all linetypes           [▼]  □ Invert filter │   └───────────┘  └────────┘
└──────────────────────────────────────────┘   ┌ Current ┐  ┌Show details┐

Current Linetype:  Continuous
┌───────────────┬──────────────┬──────────────────────────────┐
│ Linetype      │ Appearance   │ Description                   │
│ ByLayer       │  ─────────   │                              │
│ ByBlock       │  ─────────   │                              │
│ Continuous    │  ─────────── │ Continuous                   │
│ HIDDEN        │  ── ── ── ── │ Hidden __ __ __ __ __ __ __ __│
└───────────────┴──────────────┴──────────────────────────────┘
```

Press *OK* to exit the *Linetype Manager* dialogue box. The *Hidden* linetype will appear in the *Object Properties* toolbar.

Draw a series of lines on your screen to view the *Hidden* linetype.

```
┌─────────────────────────┐
│ ── ── ── HIDDEN     [▼] │
└─────────────────────────┘
```

Linetype scale

You can change the scale factor of selected linetypes at any time.

1. At the command prompt *type Chprop.*

2. Select Objects: *pick entities.*

3. Enter property to change [Color/LAyer/LType/ltScale/LWeight/Thickness]: *enter* S.

Colour selection

When you start a *New* drawing, you will see in the *Colour Control* box in the *Object Properties* toolbar that the colour is set to *Bylayer*, that is, the colour set to Layer 0, i.e. White.

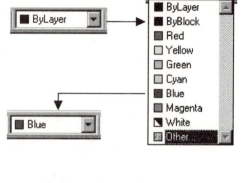

You can change this to any colour at any time irrespective of the colour setting on the current layer by selecting the *Select Colour* dialogue box.

Selection of a colour will make it current and override the colour that is set to the current layer.

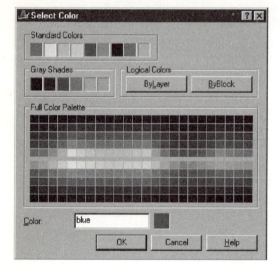

You can select from a broader range of colours by activating the *Select Colour* dialogue box by pressing *Colour...* in the *Format* pull down menu.

Editing using the Object Properties toolbar

You can quickly change an object's Layer, Colour or Linetype by using the *Object Properties* toolbar.

If you select an object outside of a command, that is, when no command is currently active, the object will be highlighted and small blue squares called *Grips* will appear.

For instance, if you selected a line, it would highlight and blue Grips would appear, one at each end and one in the middle.

The *Grips* feature will be covered at length in Chapter 11.

Whilst an object is highlighted and its Grips visible you can change its properties by selecting the required property box in the *Object Properties* toolbar.

Example

1.   Start a *New* drawing, load the linetype *Hidden* and create a layer called *Construction*.
2.   Draw a single line across the screen.
3.   Note the Layer, Colour and Linetype settings in the *Object Properties* tool bar.

♀ ✿ ⚿ 🖨■ 0 ▼	■ ByLayer ▼	———— ByLayer ▼

4.   Use your cursor to select the line just created. It will highlight and show Grips.
5.   Select the Layer Control box and select *Construction*. Select the Colour Control and select the colour *Blue* and finally, select the Linetype Control box and select the *Hidden* linetype.

♀ ✿ ⚿ 🖨■ Construction ▼	■ Blue ▼	– – – – – – HIDDEN ▼

6.   Press the *Escape* button twice to remove the Grips. Your line will now adopt these new characteristics.

Exercise using the layers, linetypes and colours

Start a *New* drawing and create the following layers and assign the colours and linetypes indicated.

Layer name	Layer colour	Layer linetype	Lineweight
Outline	Red	Continuous	0.30 mm
Hidden Detail	Blue	Hidden	Default
Centre Lines	Magenta	Center	Default

Produce the following drawing assigning the hidden detail to layer *Hidden Detail* and the centre lines to layer *Center*.

Do not dimension your drawing.

Upon completion of the drawing, *Freeze* layers *Hidden Detail* and *Centre Lines*.

Save your drawing as *Layerass1.dwg*

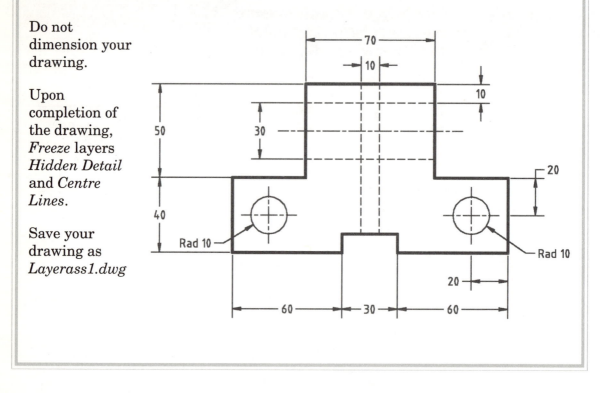

Chapter 9

Text

Textual information can be an important factor in the production of a drawing. AutoCAD has a comprehensive library of text fonts' styles from which the user can select and a variety of options to position and align the text. The following chapter will introduce you to the Dynamic text and Multiline text options, together with text editing features.

The practical assignment at the end of the chapter will build upon assignments completed in previous chapters.

Objectives

At the end of this chapter the user will be able to:

▷   Create new text styles.

▷   Place text using the DTEXT/MTEXT command, justify alignment options.

▷   Edit text in the drawing.

New commands

▷	DTEXT	DT	AⅡ	▷	DDEDIT	ED	A⁄
▷	MTEXT	MT	A	▷	SPELL	SP	

Text

You can add text to a drawing in one of two ways. AutoCAD provides the *Dtext* command for the placement of single lines of text and the *Mtext* command for paragraph text.

To create the required text style prior to entering text onto the drawing, you will need to activate the *Text Style* dialogue box by selecting *Text Style…* from the *Format* pull-down menu.

The default text style is called *Standard*. The following example shows how to create a new text style.

Example

1.   Press the *New…* button to activate the *New Text Style* dialogue box.

2. Enter a new Style called Arial then press *OK*.

3. Select the *Font Name*: edit box, scroll up the list of fonts and select *Arial*. Leave the Width factor at 1.000 and Oblique Angle at 0. Arial is now the current text style.

4. Press the *Apply* button then *Close*.

New Text Style

Style Name: Arial

OK

Cancel

Font Name:

txt.shx

AMGDT
Arial
Arial Black
Arial Narrow
BankGothic Lt BT
BankGothic Md BT

Apply

Close

Font Name:

Arial

Width Factor: 1.0000

Oblique Angle: 0

Preview

AaBbCcD

AaBbCcD   Preview

Dtext command

Draw toolbar	A
Draw menu	**Text: Single line text**
Command line entry	**DTEXT**
Alias	**DT**

The *Dtext* command displays text on screen as it is entered, you can enter multiple lines of text using *Dtext*, and you can use the Backspace key to edit the text. The text that you create will have a style, can be made any size, and can also be rotated and justified about its insertion point. The *Dtext* command will only allow you to use text styles created using the *Text Style* dialogue box.

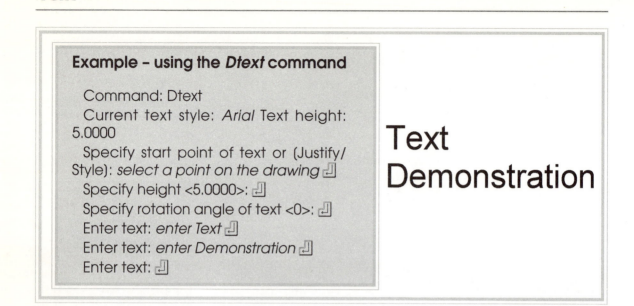

Example – using the *Dtext* command

Command: Dtext
Current text style: *Arial* Text height: 5.0000
Specify start point of text or (Justify/Style): *select a point on the drawing* ↵
Specify height <5.0000>: ↵
Specify rotation angle of text <0>: ↵
Enter text: *enter Text* ↵
Enter text: *enter Demonstration* ↵
Enter text: ↵

Text
Demonstration

Mtext command

Draw toolbar	**A**
Draw menu	**Text:**
	Single line text
Command line entry	**MTEXT**
Alias	**MT**

The *Mtext* command creates paragraphs that fit within a text boundary. The boundary defines the width of the paragraph to be entered onto the drawing. *Mtext* also allows you to specify text justification, style, height, rotation, width, colour, spacing, and other text attributes. You can select any Font within the Character tab, irrespective of the text styles created using the *Text Style* dialogue box for use with the *Dtext* command.

Example – using the *Mtext* command

Command: Mtext
Specify first corner: *select a point on your screen*
Specify opposite corner or (Height/Justify/Line spacing/Rotation/Style/Width): *drag the diagonal corner to create a rectangle.*

First Corner

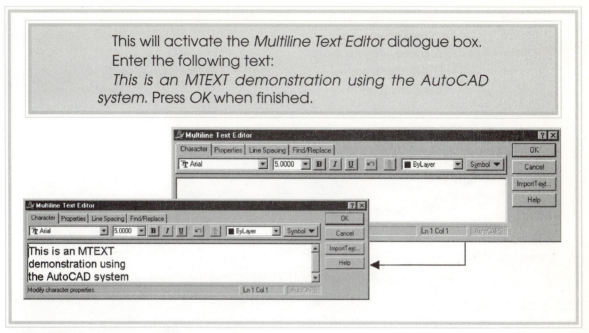

Multiline text editor tabs

The character tab

The character tab provides control of font, text height, bold, italic, underscore, stacked text, undo, colour and insertion of special characters.

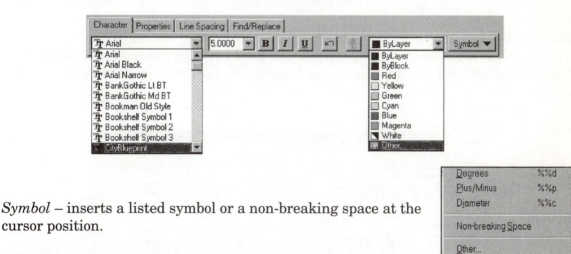

Symbol – inserts a listed symbol or a non-breaking space at the cursor position.

Using control codes

AutoCAD text can be enhanced with the addition of special characters such as the degree symbol, plus/minus tolerance symbol, and the diameter symbol. You can place these characters by using control codes.

Code	Purpose	Input Example	Output
%%d	Degree symbol (°)	25.48%%d	25.48°
%%p	Tolerance symbol (±)	25.48%%p.01	25.48±.01
%%c	Diameter symbol (ø)	%%c25.48	ø25.48
%%%	Percentage symbol (%)	25.48%%%	25.48%

Example

1. Activate the Multiline Text Editor and enter the following:

Angle between faces = 45

2. Press the *Symbol* pull down menu and select Degrees,

3. The degree symbol will appear at the cursor position.

Symbol ▼ ———► Degrees ---- %%d

Angle between faces = 45°

Choosing *Other* from the list displays the *Unicode Character Map* dialogue box showing the entire character set of the current font. To insert a character from the dialogue box, select it, and then copy and paste it into the *Multiline Text Editor* dialogue box.

Properties tab

Controls properties that apply to the paragraph object.

These include the text style, justification, width of *Mtext* box and text rotation.

| Character | Properties | Line Spacing | Find/Replace |

Style: Arial   Justification: Top Left   TL   Width: 70.6294   Rotation: 0

Top Left	TL
Middle Left	ML
Bottom Left	BL
Top Center	TC
Middle Center	MC
Bottom Center	BC
Top Right	TR
Middle Right	MR
Bottom Right	BR

Rotation values: 0, 15, 30, 45, 60, 75, 90, 105, 120

Line spacing tab

Controls the line spacing properties of the text.

| Character | Properties | Line Spacing | Find/Replace |

Line spacing: At Least   Single (1.0x)

Find/Replace tab

Searches for specified text strings and replaces them with new text.

| Character | Properties | Line Spacing | Find/Replace |

Find: [ ]   Replace with: [ ]   ☐ Match Case   ☐ Whole Word

Exercise 1 – creating text styles

Use the *Text Style* dialogue box to create the following text styles

Style Name	Font	Height	Width Factor	Oblique Angle
Times	Times New Roman	10	1	0
Country	Country Blueprint	8	0.8	0
Comic	Comic Sans MS	10	1	0
ISO	Isocp.shx	5	0.75	30

Using the *Dtext* command, create the following lines of text:

Example of Times text style

Example of Country text style

Example of Comic text style

Example of ISO text style

Exercise – using the Mtext command

Use the *Mtext* command to create the following paragraph of text.

Make your *Mtext* text boundary 200 units by 50 units. Set your text style to *Times New Roman*. Set Height to 8.

The MTEXT command enables the user to create or import large sections of text within the AutoCAD drawing.

Text editing

Modify II toolbar	A⃗
Modify menu	**Text...**
Command line entry	**DDEDIT**
Alias	**ED**

Using the *Ddedit* command

Use the *Ddedit* command to alter an existing *Dtext* or *Mtext* text string.

Example 1

Activate the *Ddedit* command
Command: ddedit
Select an annotation object or (Undo):
Select the Times text string indicated.
This will activate the *Edit Text* dialogue box
Change the *Text* string to:
Modified Times style text string
Press *OK* to close dialogue box then *Return* to exit the command

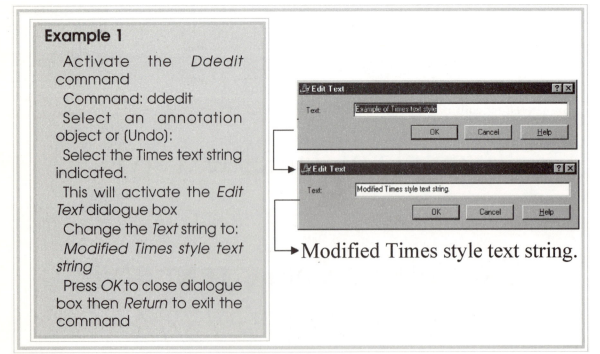

Modified Times style text string.

Example 2

Activate the *Ddedit* command
Command: Ddedit
Select an annotation object or (Undo):
select the MTEXT block of text indicated.
Modify the *MTEXT* string to:
The DDEDIT command enables the user to alter MTEXT within the AutoCAD drawing
Press *OK* to close dialogue box then *Return* to exit the command.

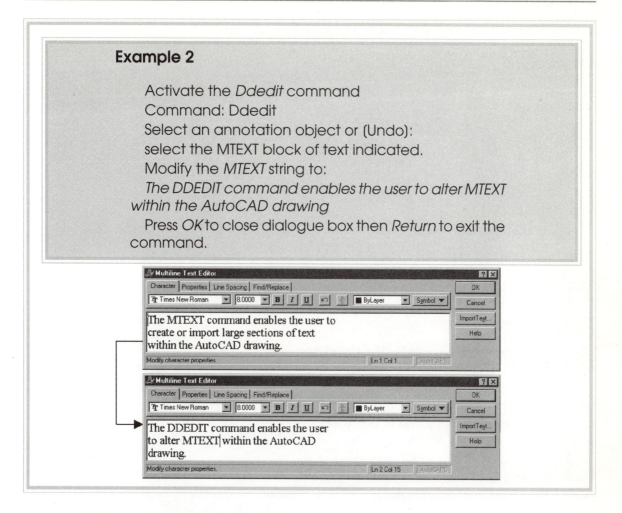

Correcting spelling mistakes

Tools menu	**Spelling**
Command line entry	**SPELL**
Alias	**SP**

Spell command

Text Objects can be checked using the *Spell* command. A selected *Mtext* or *Dtext* mis-spelt text will activate the following dialogue box.

Example

1. Activate the Spelling command
2. Select any part of the *Mtext* paragraph. This will activate the *Check Spelling* dialogue box.

Text Alignment

When using the *Dtext* command the default alignment option is bottom left of text.

You are prompted for a text height if your current text style height value is set to zero, and rotation value.

Insertion point

There are a variety of text alignment options available to enable you to place text more precisely.

When you activate the *Dtext* command you will be presented with the following command line options:

Specify start point of text or [Justify / Style]:

Enter *J* to activate the text *Justify* options.

Enter an option: [Align / Fit / Center / Middle / Right / TL / TC / TR / ML / MC / MR / BL / BC / BR]:

The *Justify* Option controls the following text alignments.

Align

Specifies both text height and text orientation by designating the endpoints of the baseline. The text will be fitted between the two points, the text height reducing as the text string becomes longer.

Insertion point

144

Fit

Fits text within an area and at an orientation defined by two points.

Fit Justify
Insertion point
2

Centre

Aligns text from the horizontal centre of the baseline, which you specify with a point.

Centre Justify
Insertion point

Middle

Aligns text at the horizontal centre of the baseline and the vertical centre of the height you specify with a point.

Middle Justify
Insertion point

Right

Specifies the right endpoint of text baseline.

Right Justify
Insertion point

As well as the text placement options discussed, you can specify justification based upon the 9 locations.

Practical assignment 3

This assignment will further develop the drawing created in Practical assignment 2 by adding layers, colours and text to the drawing.

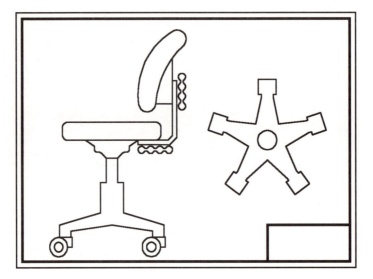

1. *Open* drawing file *Swivel_chair*.

Create the following layers. Assign the names and colours indicated.

Dimensions	Blue
Text	Red
Border	Magenta
Notes	Cyan
Chair	White

2. *Zoom* into the lower right corner of the drawing.

3. Draw a *Line* from co-ordinates 215,25. Use Object Snap *Perpendicular* to end the Line at the border edge.

4. Use command *Offset*, set distance 10 and create the second Line.

215,25

5. Create the following Text Styles.

Style Name	Font	Height	Width factor	Oblique Angle
Arial	Arial	0	1	0
Iso	Isocp.shx	0	.75	0
Simplex	Simplex	0	0.8	0

6. Set *Simplex* as the current text style and set text height to 3.

Enter *Title:, Drawn By:, Scale: and Date:* in the appropriate boxes similar to that indicated.

Title:

Drawn By:

Scale: Date:

7. Set *Arial* as the current text style and set text height to 4.

Enter date, the scale 1:10, your name and the title *Swivel Chair* in the appropriate boxes.

8. Select each of the Arial text placed in the title box to activate their respective grips.

Title: **Swivel Chair**

Drawn By: **A. Student**

Scale: **1:10**   Date: **Feb 01**

9. Using the *Object Properties* toolbar, move the selected text to layer *Text*. The text should turn red in colour.

10. Select the two border rectangles, the lines that make up the title box and the *Simplex* text within the title box to activate their respective grips.

11. Using the *Object Properties* toolbar, move the selected objects to layer *Border*.

12. Select all the Swivel Chair objects to activate their respective grips.

13. Using the *Object Properties* toolbar, move the selected objects to layer *Chair*.

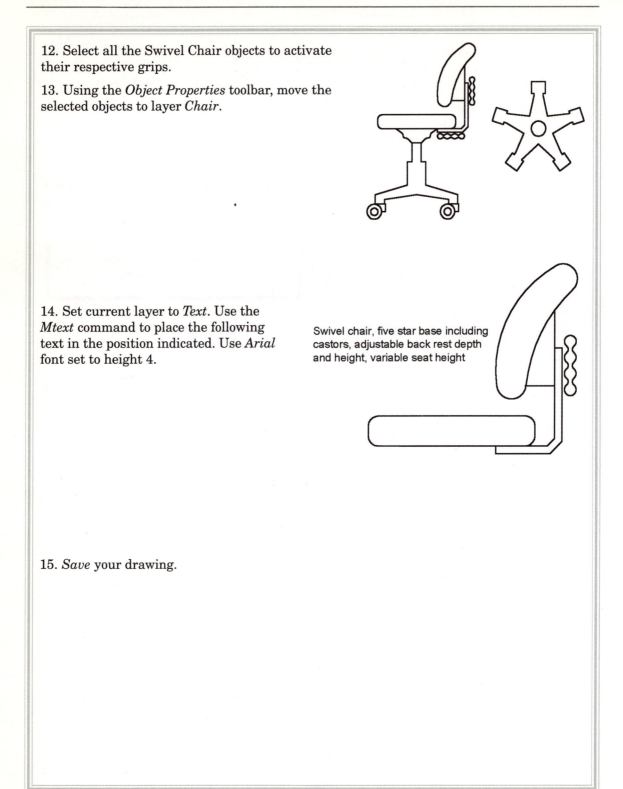

14. Set current layer to *Text*. Use the *Mtext* command to place the following text in the position indicated. Use *Arial* font set to height 4.

Swivel chair, five star base including castors, adjustable back rest depth and height, variable seat height

15. *Save* your drawing.

Multiple choice questions

1 In which toolbar would you find the Layer pull-down menu?

A Standard toolbar

B Draw toolbar

C Layers toolbar

D Object properties toolbar

2 Which of the following variables cannot be set using the Layer pull-down menu?

A Layer lock/unlock

B Layer freeze/thaw

C Layer colour

D Layer name

3 With reference to Locked layers, which of the following is true?

A You can change the colour of an object placed on a locked layer

B Objects on a Locked layer are not visible

C You cannot use the Object snap commands on any object placed on a locked layer

D Objects on this layer cannot be edited

4 Which of the following factors can you set when creating a new Text Style?

A Height, length and oblique angle

B Oblique angle, width factor and length

C Width factor, height and depth

D Width Factor, Height and Oblique Angle

5 Which of the following is true when creating a new Text Style?

A The style name must be the same as the text font

B The style name must be called *Standard*

C You do not have to set a font height

D The width factor must not exceed a value of 1

6   What is the Alias for the *Dtext* command?

A   DT

B   DX

C   DXT

D   DTXT

7   Which of the following Tabs does not appear in the Multiline Text Editor?

A   Character

B   Properties

C   Line width

D   Find/replace

8   Which of the following control codes will result in a diameter Ø symbol?

A   %%d

B   %%p

C   %%c

D   %%%

9   Which statement concerning the *Ddedit* command is true?

A   Delete selected *Mtext* text

B   Change a *Dtext* text string

C   Enable you to change the position and height of a *Dtext* entry

D   Enable you to change the text style of a *Dtext* entry

10 Which text alignment option fits text between two points but the text height will be dependent upon the size of the text string?

A   Align

B   Fit

C   Middle

D   Centre

Chapter 10

Hatching

This section investigates the various methods by which AutoCAD can hatch a drawing, and introduces the user to its comprehensive hatch pattern library. The method of selecting drawing objects for hatching is explored as are the various functions contained within the Hatching command.

Objectives

At the end of this chapter you will be able to:

▷   Be familiar with the Hatching feature.

▷   Select and scale a hatch pattern.

▷   Insert a hatch pattern.

New command

▷   Hatch   H

Hatching

Draw toolbar	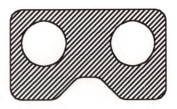
Draw menu	**Hatch...**
Command line entry	**BHATCH**
Alias	**H**

The Hatching feature enables you to create hatch patterns within closed areas of the drawing.

Hatch defines boundaries automatically when you specify a point within the area to be hatched. Any whole or partial objects that are not part of the boundary are ignored and do not affect the hatch. The boundary can have islands that you choose to hatch or leave unhatched. Islands are enclosed areas within the hatch area. You can also define a boundary by selecting objects.

The *Boundary Hatch* dialogue box is accessed from the *Draw* toolbar and contains the most common parameters used in controlling hatch pattern definition.

Quick tab features

Type

Sets the pattern type.

There are three options.

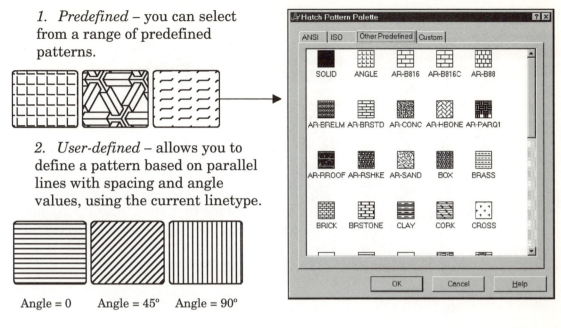

1. *Predefined* – you can select from a range of predefined patterns.

2. *User-defined* – allows you to define a pattern based on parallel lines with spacing and angle values, using the current linetype.

Angle = 0   Angle = 45°   Angle = 90°

3. *Custom* – allows you to select a hatch pattern definition not found in the supplied *acad.pat* and *acadiso.pat* files.

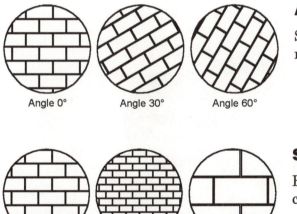

Angle 0°   Angle 30°   Angle 60°

Angle

Specifies an angle for the hatch pattern relative to the X axis.

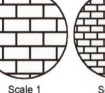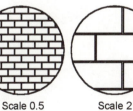

Scale 1   Scale 0.5   Scale 2

Scale

Expands or contracts a predefined or custom pattern.

Pick Point

Allows you to determine a boundary from a set of existing objects from which you want to form an enclosed area. AutoCAD prompts you for a point inside the boundary to be hatched.

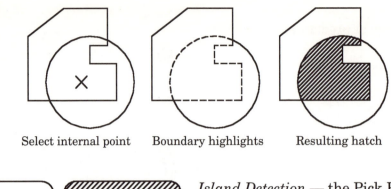

Select internal point   Boundary highlights   Resulting hatch

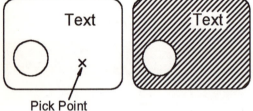

Pick Point

Island Detection — the Pick Points option will detect islands that are within the boundary generated. The space within these islands will not be hatched. This will also apply if the object(s) in question are text.

Select Objects

Only the object selected will be hatched. The hatch will ignore objects that have not been selected as boundaries.

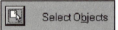

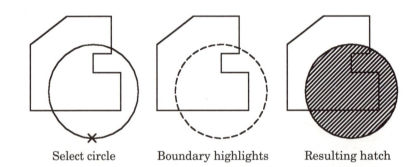

Select circle   Boundary highlights   Resulting hatch

Example – using Pick Points

1. Start a *New* drawing using *the Start From Scratch – Metric* option.
2. Create the drawing shown. Do not include dimensions.

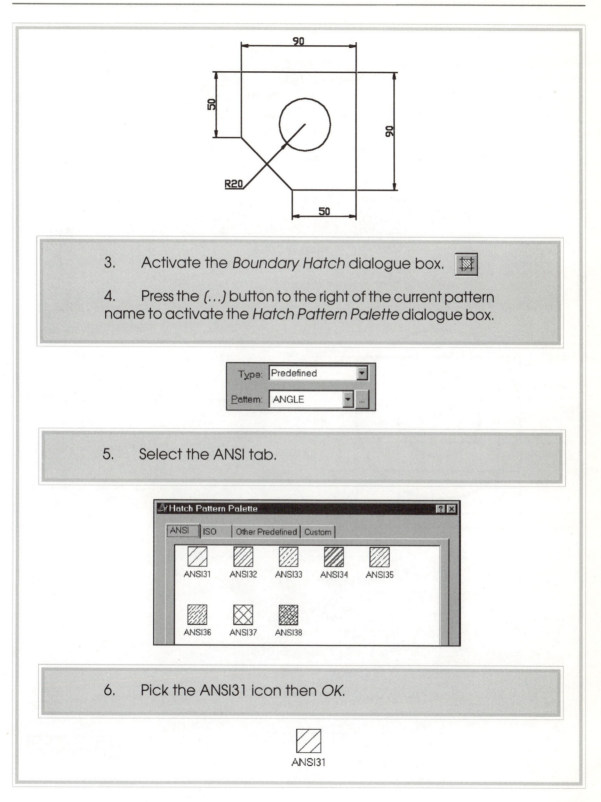

3.   Activate the *Boundary Hatch* dialogue box.

4.   Press the *(...)* button to the right of the current pattern name to activate the *Hatch Pattern Palette* dialogue box.

5.   Select the ANSI tab.

6.   Pick the ANSI31 icon then *OK*.

7. Press the *Pick Points* button.

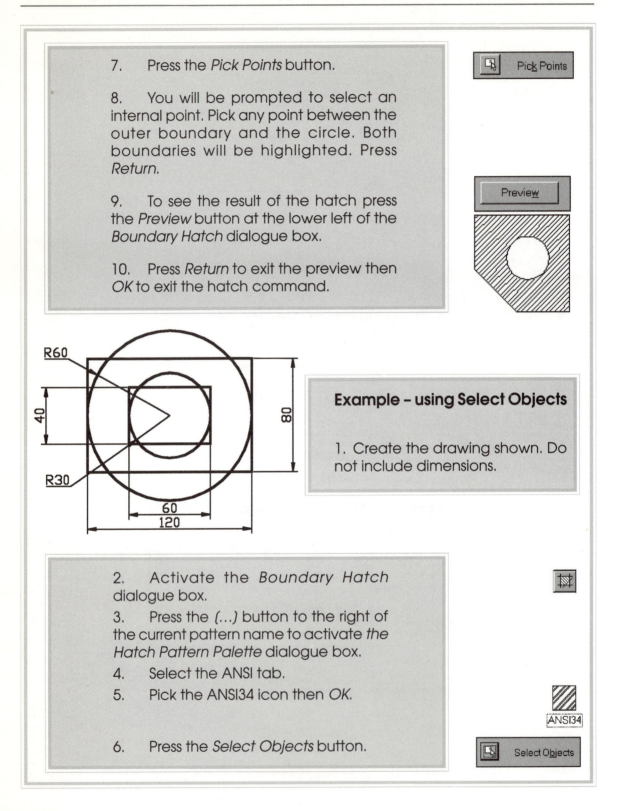

8. You will be prompted to select an internal point. Pick any point between the outer boundary and the circle. Both boundaries will be highlighted. Press *Return*.

9. To see the result of the hatch press the *Preview* button at the lower left of the *Boundary Hatch* dialogue box.

10. Press *Return* to exit the preview then *OK* to exit the hatch command.

Example – using Select Objects

1. Create the drawing shown. Do not include dimensions.

2. Activate the *Boundary Hatch* dialogue box.

3. Press the *(...)* button to the right of the current pattern name to activate *the Hatch Pattern Palette* dialogue box.

4. Select the ANSI tab.

5. Pick the ANSI34 icon then *OK*.

6. Press the *Select Objects* button.

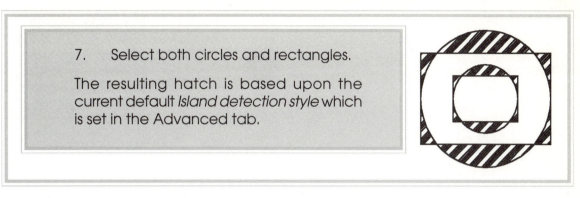

7. Select both circles and rectangles.

The resulting hatch is based upon the current default *Island detection style* which is set in the Advanced tab.

Advanced tab features

Island detection style

Island detection defines how AutoCAD specifies which boundaries are to be hatched.

There are three options:

Normal – turns on and off hatching as encounters each intersection.

Outer – only hatches between the first and second boundaries that AutoCAD encounters.

157

Ignore – ignores all internal objects and hatches through them.

Associative hatching

An associative hatch will update as its boundary is modified.

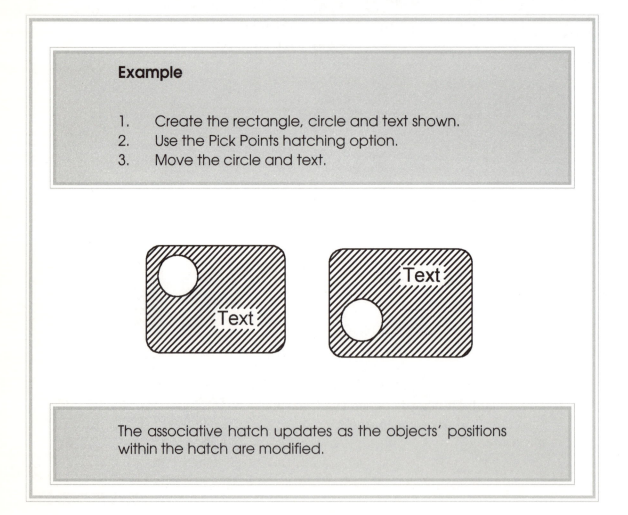

Example

1. Create the rectangle, circle and text shown.
2. Use the Pick Points hatching option.
3. Move the circle and text.

Text

Text

The associative hatch updates as the objects' positions within the hatch are modified.

Exercises – using the boundary hatch command

Create and hatch the following drawings.

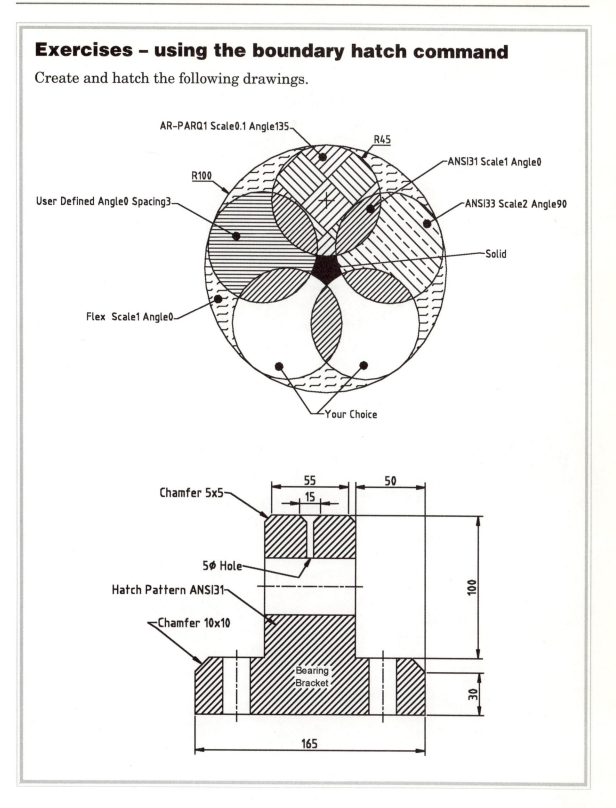

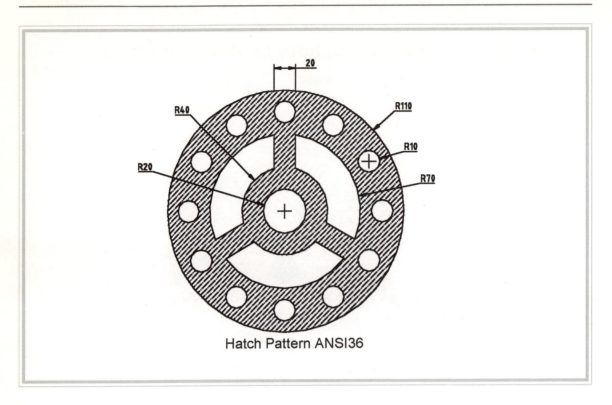

Hatch Pattern ANSI36

Chapter 11

Dimensioning

In the course of this section you will be introduced to the AutoCAD dimensioning feature. The appearance of the dimensions and the way they enter the drawing database are controlled by a set of dimensioning variables. You will learn how to manage these variables to control the appearance of dimensions with dimension styles. This dimensioning feature will enable you to dimension the most complex of drawings using a variety of styles.

Objectives:

At the end of this chapter you will be able to:

▷   Apply dimensions to linear and radial objects.
▷   Be familiar with the anatomy of a dimension.
▷   Understand and use the Dimension Style Manager dialogue box.
▷   Alter existing dimension styles.
▷   Create dimensioning styles.
▷   Complete a dimensioning assignment.

New commands

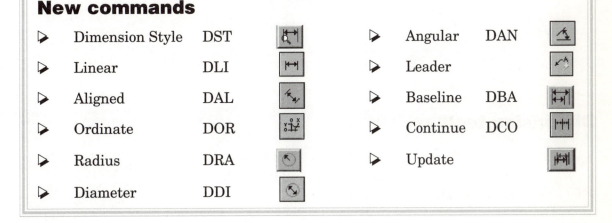

▷	Dimension Style	DST		▷	Angular	DAN	
▷	Linear	DLI		▷	Leader		
▷	Aligned	DAL		▷	Baseline	DBA	
▷	Ordinate	DOR		▷	Continue	DCO	
▷	Radius	DRA		▷	Update		
▷	Diameter	DDI					

Dimensioning

Dimensioning is one of the most important, and time consuming features of a drawing. A complex drawing may require a variety of dimensions such as Linear, Radial and Angular dimensions.

Dimensions essentially show the geometric measurements of objects in terms of Linear distances, Radial distances and Angular measurements.

Linear dimensions include horizontal, vertical, aligned, ordinate, baseline, and continue dimensions, examples of which are shown below.

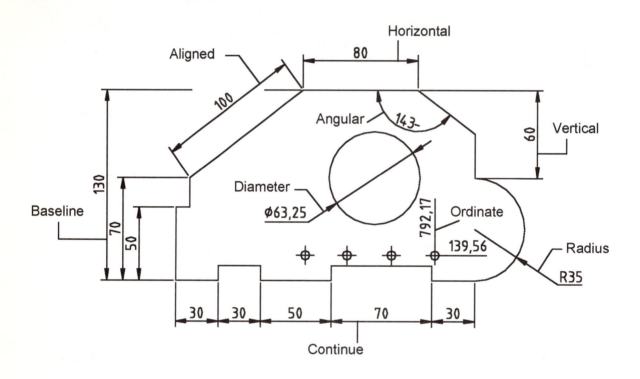

Dimension toolbar

You can activate the toolbar of your choice by right clicking any tool on any toolbar. Right click any tool in the Standard toolbar and select the Dimension toolbar from the list.

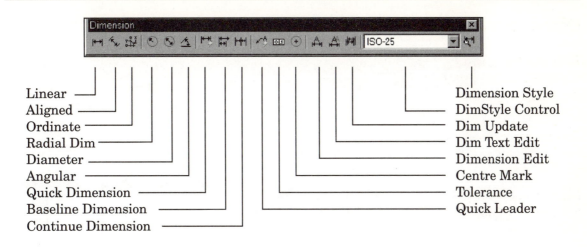

Linear
Aligned
Ordinate
Radial Dim
Diameter
Angular
Quick Dimension
Baseline Dimension
Continue Dimension

Dimension Style
DimStyle Control
Dim Update
Dim Text Edit
Dimension Edit
Centre Mark
Tolerance
Quick Leader

The following illustrations show examples of the various dimension types and associated tools.

Horizontal Dimension

60,00

Vertical Dimension

25,00

Align Dimension

20,62

Ordinate

565,00

115,00

Radial Dimension

R10,00

Diameter Dimension

Ø12,07

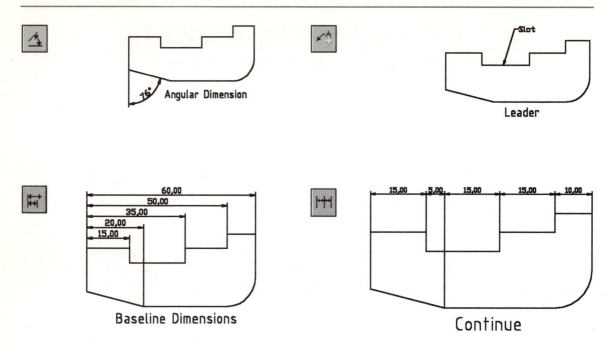

Angular Dimension

Leader

Baseline Dimensions

Continue

Anatomy of a dimension

Dimension line – A dimension line is a line that indicates the direction and extent of a dimension. For angular dimensioning, the dimension line is an arc.

Extension line – Extension lines extend from the feature to the dimension line.

Arrowhead – Arrowheads are added to each end of the dimension line.

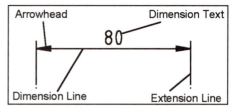

Dimension text – Dimension text is an optional text string that usually indicates the actual measurement. The text may also include prefixes, suffixes, and tolerances.

Leader lines – A leader is a solid line leading from some annotation to the referenced feature.

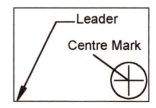

Centre mark – A centre mark is a small cross that marks the centre of a circle or arc.

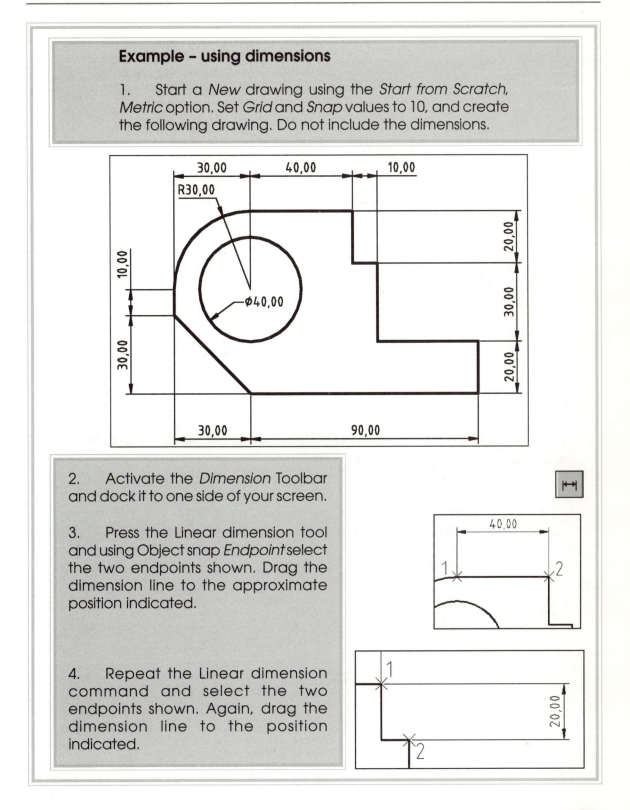

Example – using dimensions

1.   Start a *New* drawing using the *Start from Scratch, Metric* option. Set *Grid* and *Snap* values to 10, and create the following drawing. Do not include the dimensions.

2.   Activate the *Dimension* Toolbar and dock it to one side of your screen.

3.   Press the Linear dimension tool and using Object snap *Endpoint* select the two endpoints shown. Drag the dimension line to the approximate position indicated.

4.   Repeat the Linear dimension command and select the two endpoints shown. Again, drag the dimension line to the position indicated.

5. Repeat the Linear dimension command until your drawing is similar to that shown.

Note – the dimensions are created using the default ISO-25 dimension style. You will continue to use this style to create further dimensions on your drawing.

6. Press the Align dimension tool and using Object snap *Endpoint* select the two endpoints shown. Drag the dimension line to the approximate position indicated.

7.   Press the Radial tool and select the point on the profile indicated.

8.   Press the Diameter tool and place the diameter dimension in the position indicated.

9.   Finally, press the Angular tool and create the dimension indicated.

10.   Your drawing should now be similar to that shown.

11.   Save your drawing as *Dimass1*.

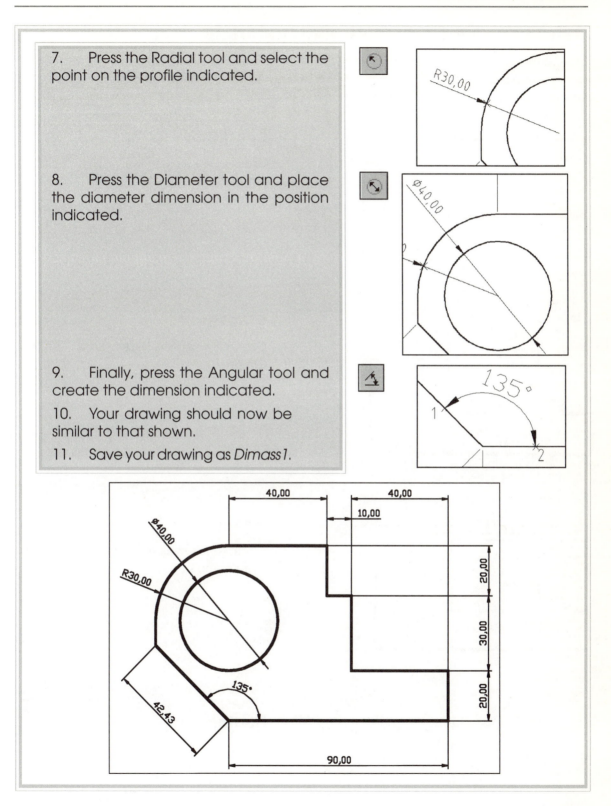

Qdim command for quick dimensioning

The *Qdim* command is used to quickly create a series of dimensions. The command reduces the time required to produce a series of baseline or continued dimensions, or for dimensioning a series of circles and arcs.

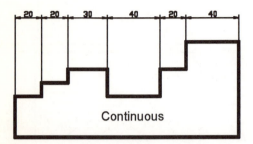

Continuous

Creates a series of continued dimensions.

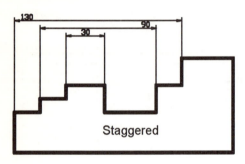

Staggered

Creates a series of staggered dimensions.

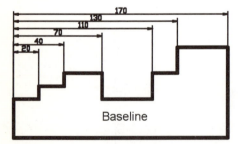

Baseline

Creates a series of baseline dimensions.

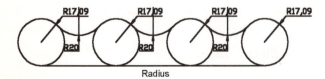

Radius

Creates a series of radius dimensions.

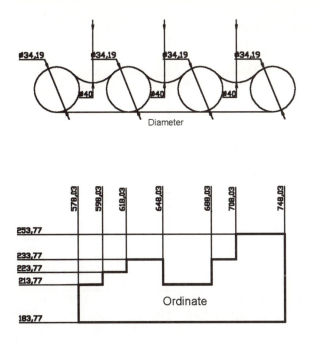

Diameter

Creates a series of diameter dimensions.

Ordinate

Creates a series of ordinate dimensions.

Exercise

Create a simple polyline shape similar to the examples shown above and experiment with the various *Qdim* options.

Associative dimensions

Dimensions are by default associative, which means that they will modify their value and position as the entity related to it is modified. To illustrate this draw a rectangle, then dimension one of its sides. Use the Stretch command to change the length of the rectangle, ensuring that the object selection crossing window also selects the dimension. As the rectangle changes in length so to will its corresponding dimension value.

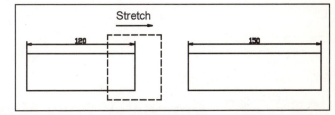

Dimension styles

Every dimension has a dimension style associated with it. You can use the default dimension style or define your own.

169

Standard   This is the default dimension style when you use the English (Feet and Inches) template.

ISO-25   This is the default dimension style when you use the Metric template.

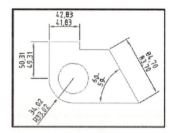

Limits   You can apply tolerances to the measurement, the dimension then consists of the high and low values of the measurement rather than just the nominal value.

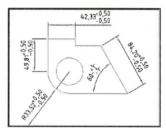

Tolerances   Dimension tolerances are plus and minus values that can be appended to the dimension text generated.

Alternate Units   Two systems of measurement can be created simultaneously, commonly used to indicate metric and imperial values.

Dimension Style Manager

The *Dimension Style Manager* dialogue box is used to modify existing or create new dimension styles. To activate this dialogue box, select *Dimension Style...* from the *Format* menu.

The Preview graphics box shows the current setting for the various linear, radial and angular dimensions. In the case of the ISO-25 style shown, this means that dimensions are set to 2 decimal places and dimensions are aligned along the dimension line using the *Standard* text style.

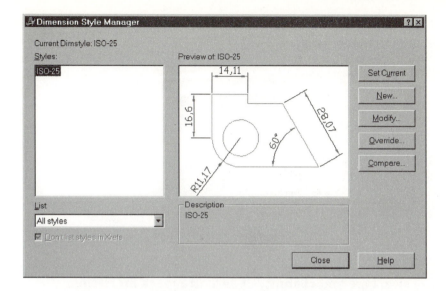

Let's look at the details of this dimension style a little more closely.

Select the *Modify...* button.

The Tabs within the *Modify Dimension Style* dialogue box indicate the variables that together determine the dimension style. Each tab contains a graphics window which shows the result of current settings.

Lines and arrows tab

- This tab is used to set the colour and thickness of the dimension and extension lines.

- The size and type of Arrowheads, selectable from a list, used for the dimension line and Leader.

- The size, type and visibility of a Centre mark for circles.

Text tab

- Use this tab to set text style, colour and position on the dimension line
- Text alignment options include:
  - Forced Horizontal
  - Aligned with the dimension line
  - Set to the ISO standard.

Fit tab

- Use this tab to set the placement of the dimension text, line and arrows between the extension lines. If there is limited room between the extension lines, these options allow you to determine whether the text or arrows are forced outside the extension lines.

- Sets text placement when it is not possible to accept the default text position.

- Overall scaling factor. Used to scale all the elements of the dimension to suit the drawing scale.

Example

Dimension overall scale factor set to 1.

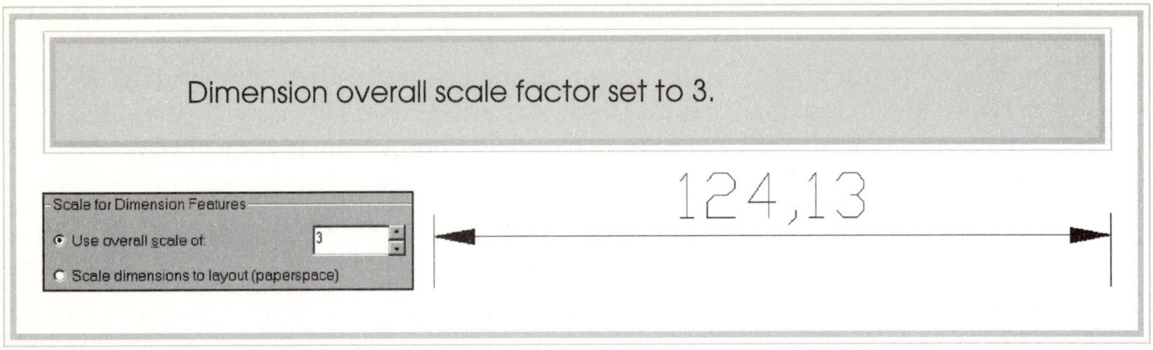

Dimension overall scale factor set to 3.

124,13

Scale for Dimension Features

○ Use overall scale of: 3

○ Scale dimensions to layout (paperspace)

Primary units tab

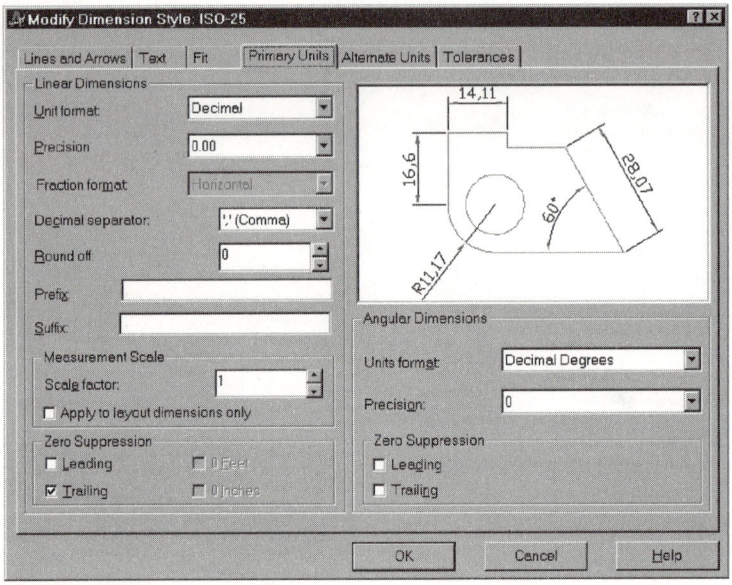

- Used to set dimension units and number of decimal places.
- Dimension text prefix/suffix.
- Zero suppression, leading or trailing.
- Units and precision of Angular dimensions.

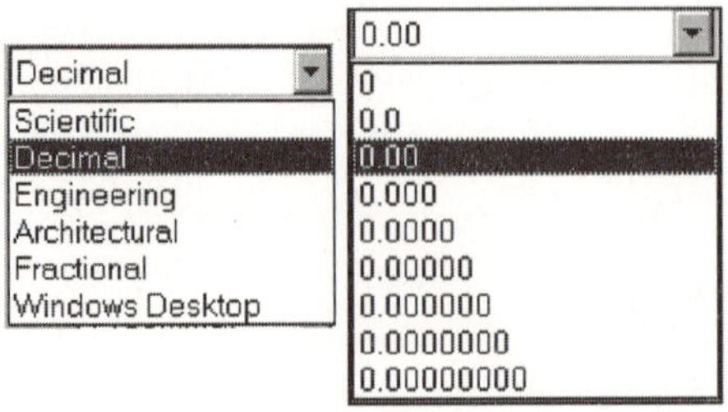

Alternate units tab

● Used to set Alternative units format and placement. Dimension text will include both Metric and Imperial equivalent dimensions.

Tolerances tab

● Use this tab to select and set one of three tolerance formats. These include:

Example

The following tolerance examples are based on a tolerance value of 0.5.
The *Basic* option places a rectangle round the dimension text.

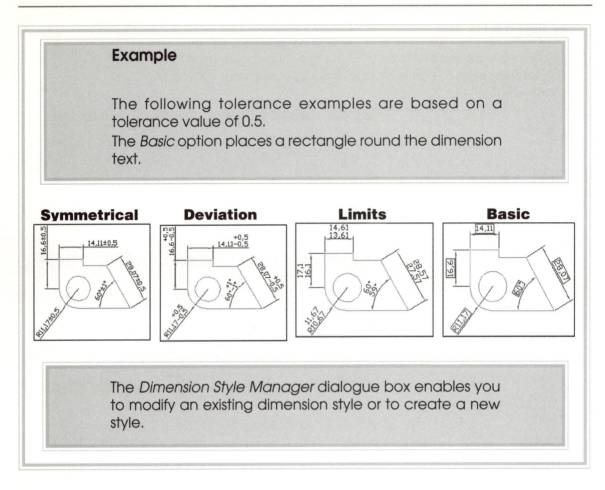

| Symmetrical | Deviation | Limits | Basic |

The *Dimension Style Manager* dialogue box enables you to modify an existing dimension style or to create a new style.

Modifying dimension styles

You can use your drawing *Dimass1* to illustrate the technique used to modify an existing dimension style.

Example – setting a radial dimension style

1.   Open the drawing file *Dimass1*.

The current default dimension style (ISO-25) is set to align dimensions along the dimension line. Whilst this is suitable for the horizontal and vertical dimensions, it is not appropriate for the radial and diameter dimensions.

The next task is to modify the current dimension style so that radial and diameter dimensions are forced horizontal.

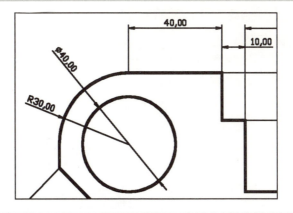

2.   Activate the *Dimension Style Manager* dialogue box.

3.   Select ISO-25, it will turn blue then press the *New...* button to activate the *Create New Dimension Style* dialogue box.

At this point we do not wish to create a new style, we only want to alter the existing settings for the Radial and Diameter dimensions.

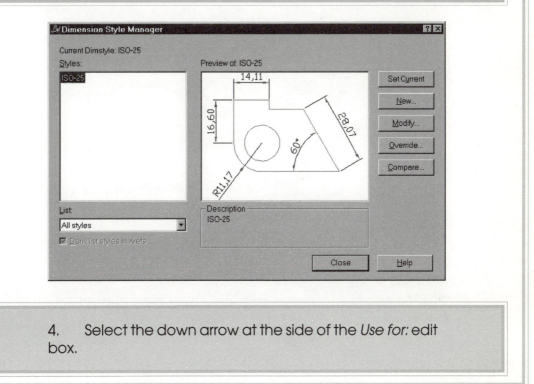

4.   Select the down arrow at the side of the *Use for:* edit box.

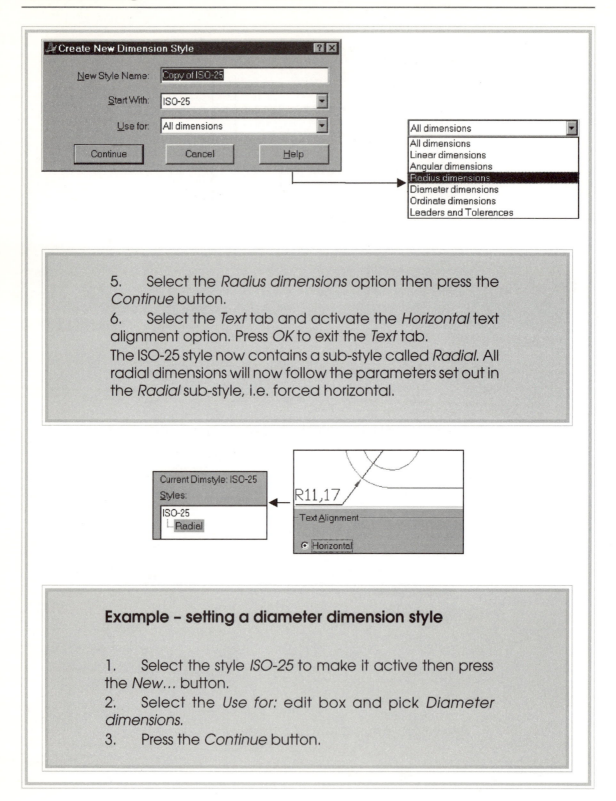

5.   Select the *Radius dimensions* option then press the *Continue* button.

6.   Select the *Text* tab and activate the *Horizontal* text alignment option. Press *OK* to exit the *Text* tab.

The ISO-25 style now contains a sub-style called *Radial*. All radial dimensions will now follow the parameters set out in the *Radial* sub-style, i.e. forced horizontal.

Example – setting a diameter dimension style

1.   Select the style *ISO-25* to make it active then press the *New...* button.

2.   Select the *Use for:* edit box and pick *Diameter dimensions*.

3.   Press the *Continue* button.

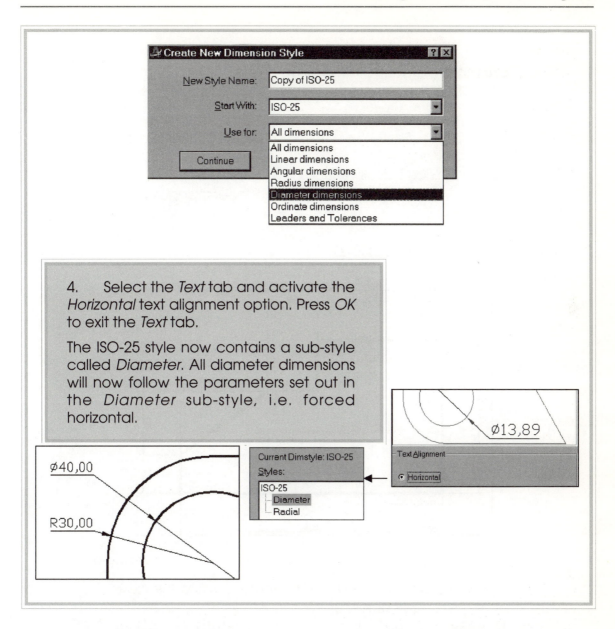

4.   Select the *Text* tab and activate the *Horizontal* text alignment option. Press *OK* to exit the *Text* tab.

The ISO-25 style now contains a sub-style called *Diameter*. All diameter dimensions will now follow the parameters set out in the *Diameter* sub-style, i.e. forced horizontal.

Creating new dimension styles

In the course of the following example, you will create two new dimension styles called *Limits* and *Alternate*.

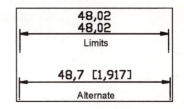

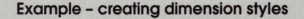

Example – creating dimension styles

1.   Start a *New* drawing using the *Start from Scratch, Metric* option. Set Grid and Snap values to 10, and create the following drawing.
Do not include the dimensions.
2.   Activate the *Dimension Style Manager* dialogue box.

Note – the current default style name is ISO-25.

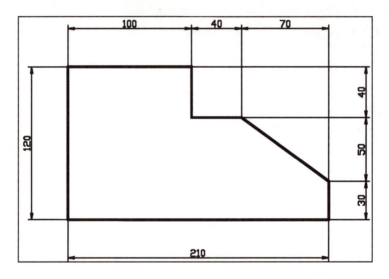

3.   Press the *New...* button and enter *Limits* in the *New Style Name*: edit box, then *Continue*.
The new style *Limits* is currently identical to the *ISO-25* dimension style.
4.   Select the *Tolerances* tab from the *New Dimension Style* dialogue box.
5.   Change the Tolerance Format Method: from *None* to *Limits*.

The graphics window will now illustrate the *Limits* format.

6. Enter the Upper and Lower values indicated below.

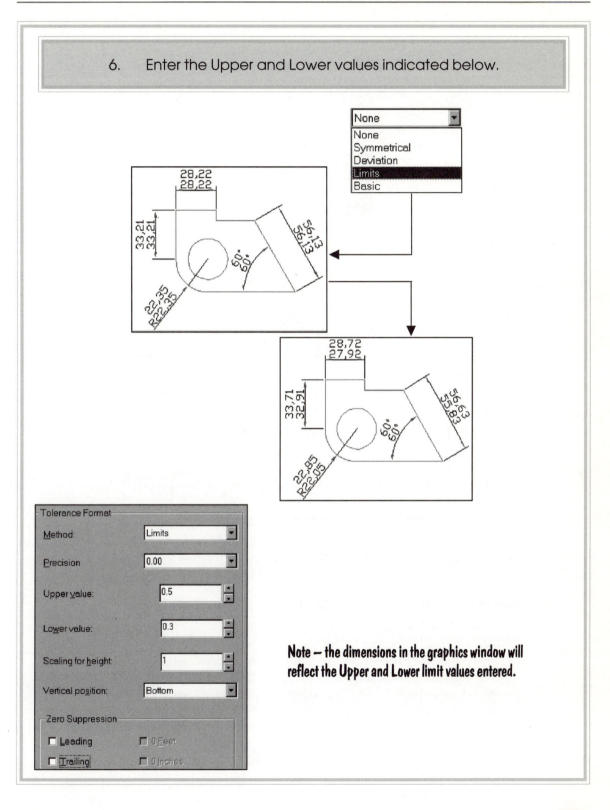

Note – the dimensions in the graphics window will
reflect the Upper and Lower limit values entered.

7.   Press the *OK* button. The new dimension style *Limits* is set but not current.

8.   Select *Limits* from the styles list then press the *Set Current* button.

9.   Finally, press the *Close* button.

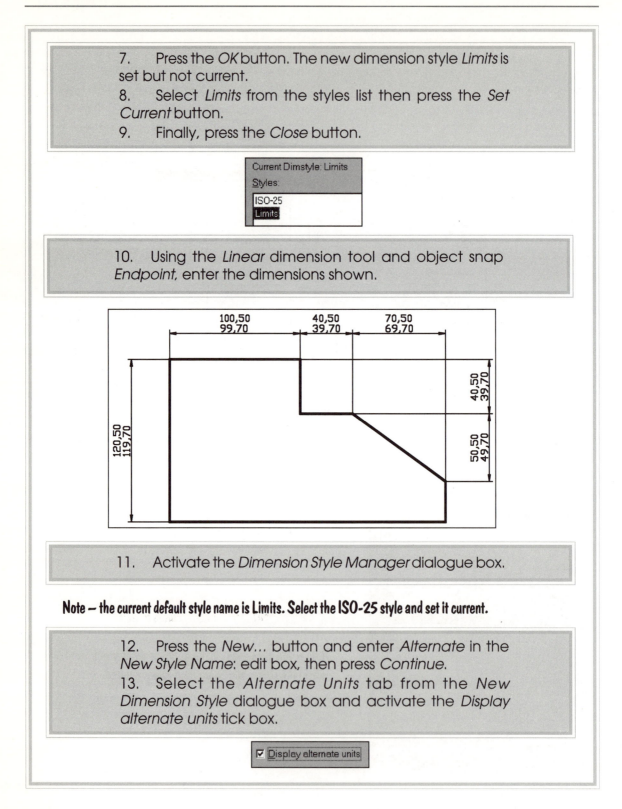

10.   Using the *Linear* dimension tool and object snap *Endpoint,* enter the dimensions shown.

11.   Activate the *Dimension Style Manager* dialogue box.

Note – the current default style name is Limits. Select the ISO-25 style and set it current.

12.   Press the *New...* button and enter *Alternate* in the *New Style Name*: edit box, then press *Continue*.

13.   Select the *Alternate Units* tab from the *New Dimension Style* dialogue box and activate the *Display alternate units* tick box.

14.   Ensure the following values are entered. This will result in dimensions in millimetres being followed by their imperial equivalent dimension in inches.

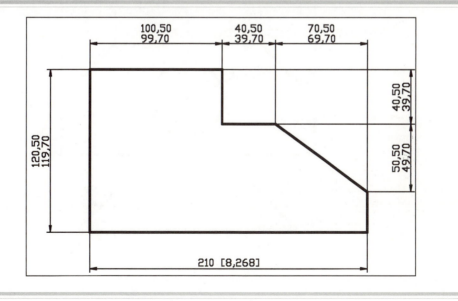

Alternate Units

Unit format	Decimal
Precision	0.000
Multiplier for alt units:	0.03937
Round distances to:	0
Prefix	
Suffix	

Zero Suppression

☐ Leading   ☑ 0 Feet
☐ Trailing   ☑ 0 Inches

15.   Press *OK* to exit the *Alternate units* tab. Set the dimension style *Alternate* current.
16.   Using the *Linear* dimension tool and object snap *Endpoint*, enter the dimensions shown.
17.   *Save* your drawing as *Dimass2.dwg*.

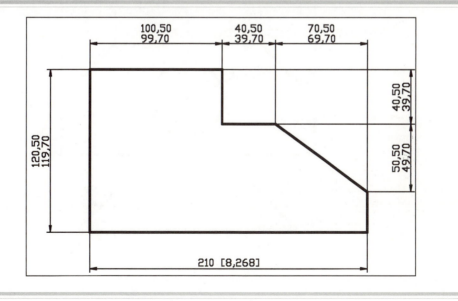

Practical assignment 4

Create and dimension as indicated the drawings illustrated below.

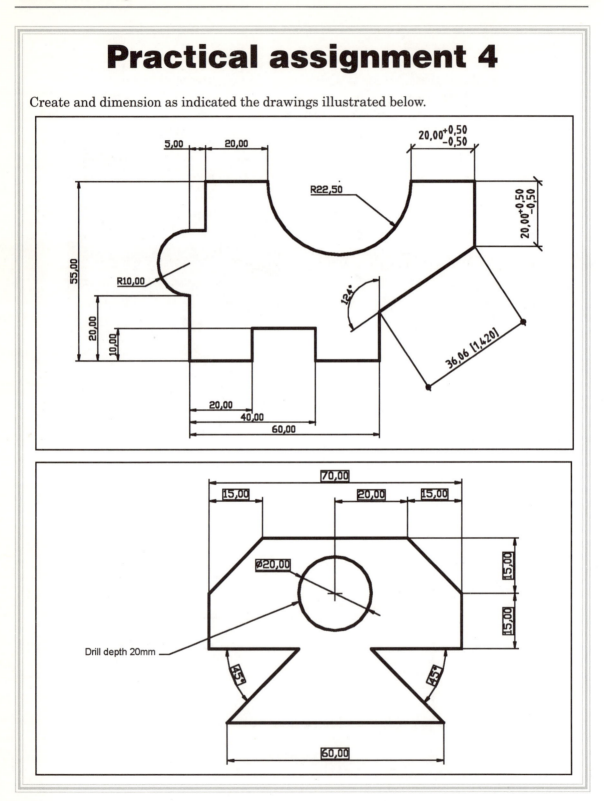

Multiple choice questions

1   Which tolerance dimension style does the following illustration represent?

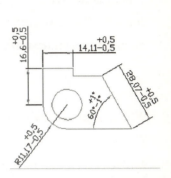

A   Basic

B   Symmetrical

C   Deviation

D   Limits

2.   Which dimension style does the following illustration represent?

A   Basic

B   Limits

C   Alternate units

D   None of the above

3.   Which tab in the Modify Dimension Style dialogue box would be used to set the decimal precision of the dimension?

A   Text fit

B   Lines and arrows

C   Tolerance

D   Primary units

4.   Which tab in the Modify Dimension Style dialogue box would you find the text alignment options?

A   Text

B   Fit

C   Primary units

D   Alternate units

5   Which of the following statements is true?

A   The size of the dimension line arrowheads cannot be altered

B   One or both extension lines can be suppressed if required

C   Only the Standard text style can be used within a dimension

D   Maximum primary units precision is six decimal places

6   Which of the following statements is true?

A   The overall scale factor only applies to linear dimensions

B   Any text style can be applied to a dimension style

C   The Standard dimension styles cannot be modified

D   Only a total of five dimension styles are allowed per drawing

7   Setting a basic dimension is done in the:

A   Primary units tab

B   Alternate units tab

C   Text tab

D   Tolerances tab

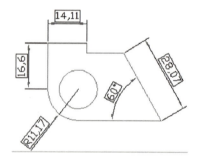

8   Dimension styles are displayed in the:

A   Modify dimension style dialogue box

B   Dimension style manager dialogue box

C   Text style dialogue box

D   None of the above

Chapter 12

Templates

Template files are pre-prepared files that contain all the elements that would be common to a set of drawings. Each time you start a new drawing, you need to create new text styles, dimension styles, layer names etc. which may be common to all your drawings. A template file is a predefined file that contains all your common settings which helps standardise the drafting process. The following chapter will take you through the process of creating a template file followed by a practical assignment that will test your understanding of the features covered.

Objectives

At the end of this chapter you will be able to:

▷   Understand the concept of using Templates.

▷   Create a template file containing your user defined features.

Templates

Many of the drawings you create will have common features such as drawing limits, layer names, linetypes and text styles. Each time you create a new drawing, time is spent reproducing these features. Template drawings store AutoCAD settings that you would otherwise have to set individually and are useful for maintaining drawing standards and conventions.

A template drawing is an AutoCAD drawing file that has the extension *.dwt* and is saved automatically in the template directory.

In a template drawing file you can predefine many settings including:

- Drawing units
- Sheet size and layout
- Text and dimensioning styles
- Colours
- Types of lines
- Layer names
- View names
- UCS names
- Title blocks.

To better understand their significance and use you will now produce a standard template file that can be used for all subsequent drawing exercises in this book.

Exercise – creating a template file

Your template file will contain the following settings:

1. Units — Decimal, precision.
2. Drawing limits — 420,297.
3. Grid and snap — Set to 10.
4. Layers — 0, construction, dimensions, hidden detail, outline, text.
5. Text styles — ISO, Arial.
6. Dimension styles — Student.

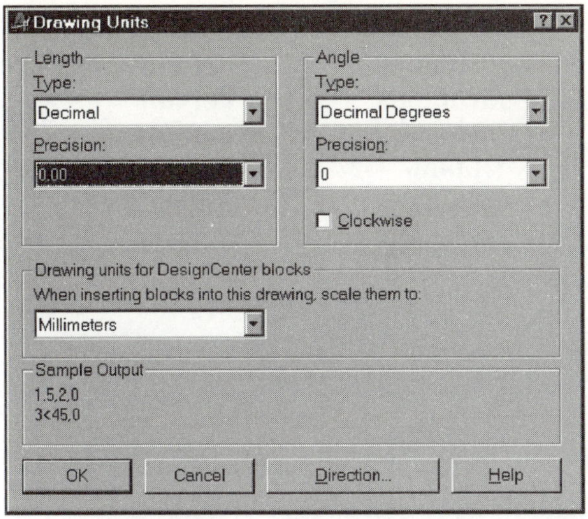
Drawing Units
Length
Type:
Decimal
Precision:
0.00
Angle
Type:
Decimal Degrees
Precision:
0
Clockwise
Drawing units for DesignCenter blocks
When inserting blocks into this drawing, scale them to:
Millimeters
Sample Output
1.5,2.0
3<45,0
OK
Cancel
Direction...
Help

Units

1. Start a *New* drawing. Select the *Start from Scratch – Metric* option.

2. From the *Format* menu select *Units...*

3. Set *Units* type to *Decimal* and Precision to 0.00.

Press *OK* to exit.

South East Essex College
of Arts & Technology

Drawing limits

1. From the *Format* menu select *Units...*

2. Specify lower left corner 0.00,0.00.

3. Specify upper right corner 420.00,297.00.

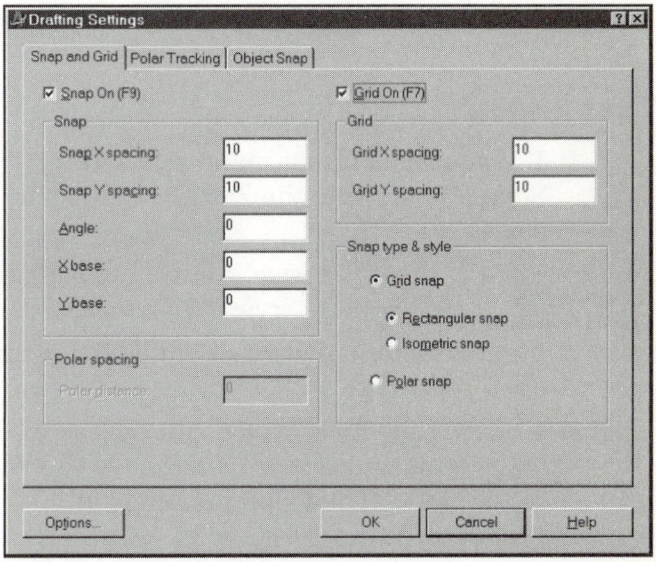
Drafting Settings
Snap and Grid
Polar Tracking
Object Snap
Snap On (F9)
Grid On (F7)
Snap
Snap X spacing:
10
Snap Y spacing:
10
Angle:
0
X base:
0
Y base:
0
Grid
Grid X spacing:
10
Grid Y spacing:
10
Snap type & style
Grid snap
Rectangular snap
Isometric snap
Polar snap
Polar spacing
Polar distance:
0
Options...
OK
Cancel
Help

Grid and Snap

1. From the *Tools* menu select *Drafting Settings...*

2. Select the *Snap and Grid* tab.

3. Set Snap X and Y to 10.

4. Set Grid X and Y to 10.

Layers

1. Activate the *Layer Properties Manager* dialogue box.

2. Create the following layers and assign the colours, linetypes and lineweights indicated.

Layer Name	Layer Colour	Layer Linetype	Lineweight
0	White	Continuous	default
Construction	Magenta	Continuous	default
Dimensions	Red	Continuous	default
Hidden Detail	White	Hidden	default
Outline	White	Continuous	0,30
Text	Blue	Continuous	default

Text styles

1. From the Format menu select *Text Style...* to activate the *Text Style* dialogue box.

2. Press the *New...* button to activate the *New Text Style* dialogue box.

3. Enter a new Style called *Arial* then press *OK*.

4. Select the *Font Name*: edit box, scroll up the list of fonts and select *Arial*. Leave the Width factor at 1.000 and Oblique Angle at 0.

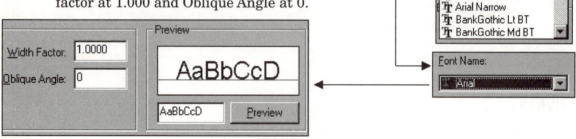

5. Press the *Apply* button.

6.   Press the *New...*
button to activate the
New Text Style dialogue
box Enter a new Style
called *ISO* then *OK*.

7.   Select the *Font Name*: *Isocp.shx*. Set width factor 0.75.

8.   *Apply* and *Close* the *Text Style* dialogue box.

Dimension style

Your first task is to create your personal dimension style copied from the current
ISO-25 dimension style.

1.   Select *Dimension Style...* from the *Format* menu to activate the *Dimension
Style* dialogue box.

2.   Press the *New...* button, enter
Student in the *New Style Name*: edit
box then press *Continue*.

3.   The *Student* dimension style is
currently identical to the *ISO-25*
dimension style.

4.   Select the *Primary Units* tab and
set *Precision* to 0 then *OK*.

5.   Now make the *Student* dimension style current
by selecting *Student* in the *Styles* list then press the
Set Current button.

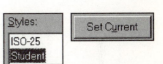

Your next task is to set radial dimensions so that they are forced horizontal.

6.   Press the *New...* button.

7.   Select the *Use for:* edit box and
pick *Radius dimensions* then press
the *Continue* button.

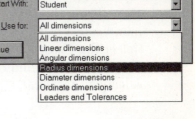

8.   Select the *Text* tab and activate the *Horizontal* text
alignment option. Press *OK* to return to the *Dimension Style
Manager*. Press *Close* to exit.

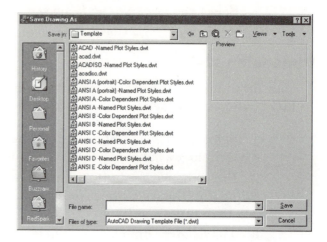

9.   Now to save the template file. From the *File* menu select *Save As...*

10.   In the *Save As Type:* edit box select: *AutoCAD Drawing Template File (*,dwt).*

11.   Enter the file name *Student* then press the *Save* button.

12.   Place general information about the template file in *the Template Description* dialogue box. Use the illustration as a guide.

13.   Press *OK*.

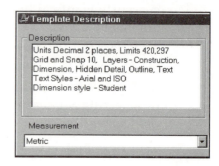

14.   From the *File* menu select *Close* to close the file.

15.   From the *File* menu select *New...* Select the *Use a Template option* and select *Student.dwt*.

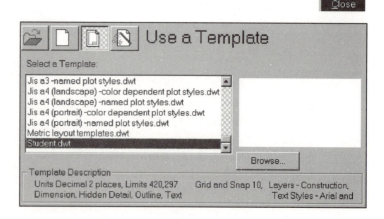

The Student template file is now current and all your settings are available for use.

Practical assignment 5

The object of this assignment is to create the following drawing using the *Student* template as the basis for the drawing:

1. Start a *New...* drawing and select the *Using a Template* option. Pick the *Student* template from the list.

2. Create the following drawing. Do not include the dimensions.

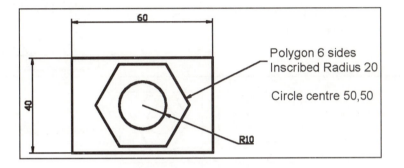

Polygon 6 sides
Inscribed Radius 20

Circle centre 50,50

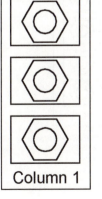

3.   Using the *Dtext* command set the text style to *Arial* and text height to 10. Enter the text *Column 1* justified *middle* below the drawing.

4.   Use the *Array* command to produce 5 rows and 1 column of the drawing. Make the vertical spacing 50 units.

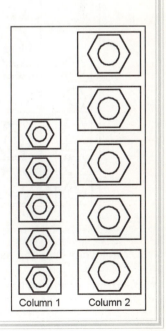

5.   *Copy* the column and text 100 units to the right then apply a *Scale* factor of 1.5 to the drawing objects, not the text below them.

6. Use the *DDedit* command to change the text *Column 1* to *Column 2*.

7. Repeat the *Copy* command and copy the second column a distance of 138 units to the right. Apply a *Scale* factor of 1.5 to the drawing objects, not the text.

8. Use the *DDedit* command to change the text *Column 2* to *Column 3*.

9. Ensure that the variable *Mirrtext* is set to 0 and *Mirror* all three columns as indicated below.

10. Use the *DDedit* command to change the mirrored text to *Column 4, Column 5 and Column 6*.

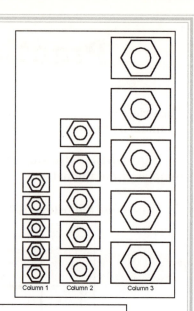

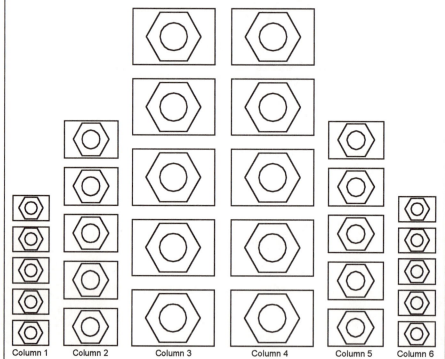

11. *Save* your drawing as *Ass5.dwg*.

This drawing will be used to illustrate the Modifying Object Properties feature in Chapter 13.

Chapter 13

Modifying object properties

Every object that has been created in a drawing has a number of properties. Taking the polyline as an example, some of its properties include linetype, linescale, colour, width, layer etc.

You can isolate objects in the drawing using the Quick Select feature and modify their object properties at any point during the creation of a drawing. In the course of this chapter you will undertake an object modifying exercise, based upon the drawing ASS5.dwg created in the previous chapter.

Objectives

At the end of this chapter you will be able to:

▷   Understand the principles of the Properties dialogue box.

▷   Apply the Properties feature to a drawing exercise.

▷   Use the Quick Select feature to identify specific objects in the drawing.

New commands

▷   Properties   

▷   Quick Select   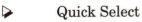

Properties command

Standard toolbar	
Tools menu	**Properties**
Command line entry	**Properties**
Alias	**PROPS**

The Properties feature allows you to change the properties of one or more objects in the drawing.

Every object has a set of properties and characteristics assigned to it.

Properties are common to all objects and can include:

Layer, Colour, Linetype and *Lineweight.*

Object characteristics are specific to the particular object, for instance, in the case of text, the characteristics that are unique to a selected text would be its contents and co-ordinate position.

Properties dialogue box

The *Properties* dialogue box is used to control selected object(s) properties.

Object properties are displayed either alphabetically or by category, depending on the tab you choose. To modify properties using the Properties window select the object whose properties you want to change.

You can identify an object for selection in two ways:

Select the object(s) to be changed then press the *Properties* tool on the *Standard* toolbar. If multiple objects are selected, the *Properties* window displays all the properties they have in common.

Alternatively, you can identify object(s) by using the *Quick Select* feature that will find the object(s) based on search data entered by the user.

We will use the drawing *Ass5.dwg* created in Practical assignment 5 to illustrate the concept of modifying object properties.

196

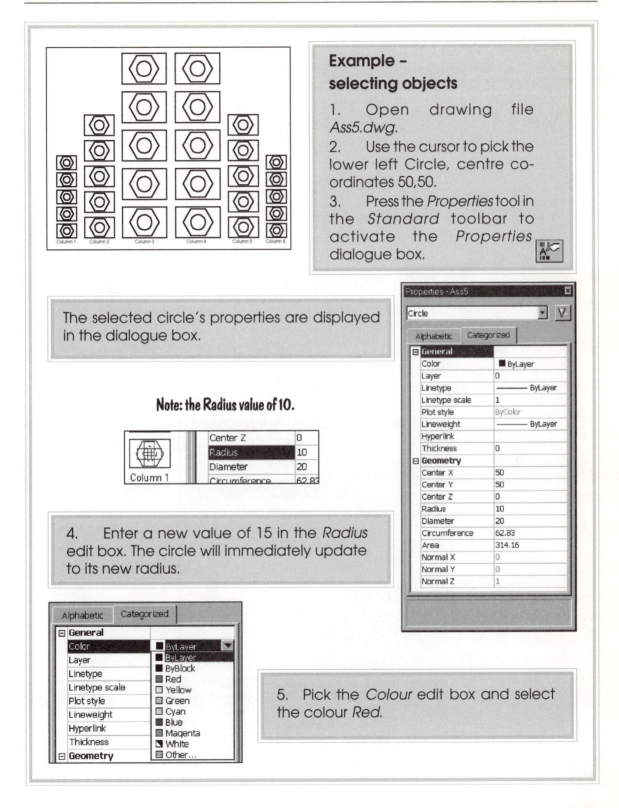

Example –
selecting objects

1. Open drawing file *Ass5.dwg.*
2. Use the cursor to pick the lower left Circle, centre co-ordinates 50,50.
3. Press the *Properties* tool in the *Standard* toolbar to activate the *Properties* dialogue box.

The selected circle's properties are displayed in the dialogue box.

Note: the Radius value of 10.

	Center Z	0
	Radius	10
	Diameter	20
Column 1	Circumference	62.83

4. Enter a new value of 15 in the *Radius* edit box. The circle will immediately update to its new radius.

Properties - Ass5

Circle

Alphabetic	Categorized	
⊟ **General**		
Color	■ ByLayer	
Layer	0	
Linetype	——— ByLayer	
Linetype scale	1	
Plot style	ByColor	
Lineweight	——— ByLayer	
Hyperlink		
Thickness	0	
⊟ **Geometry**		
Center X	50	
Center Y	50	
Center Z	0	
Radius	10	
Diameter	20	
Circumference	62.83	
Area	314.16	
Normal X	0	
Normal Y	0	
Normal Z	1	

Alphabetic	Categorized	
⊟ **General**		
Color	■ ByLayer ▼	
Layer	■ ByLayer	
Linetype	■ ByBlock	
Linetype scale	■ Red	
Plot style	□ Yellow	
Lineweight	□ Green	
Hyperlink	□ Cyan	
Thickness	■ Blue	
⊟ **Geometry**	■ Magenta	
	◣ White	
	▨ Other...	

5. Pick the *Colour* edit box and select the colour *Red.*

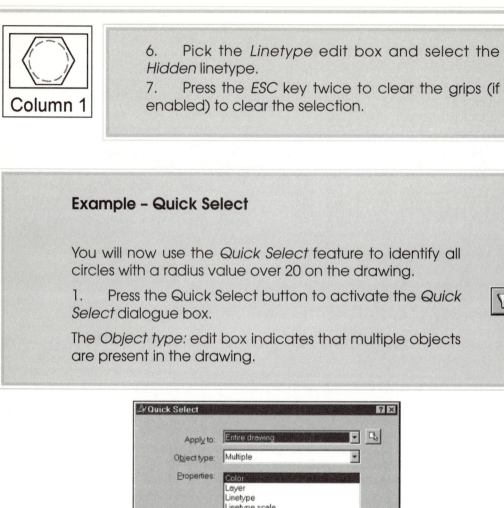

Column 1

6.   Pick the *Linetype* edit box and select the *Hidden* linetype.

7.   Press the *ESC* key twice to clear the grips (if enabled) to clear the selection.

Example – Quick Select

You will now use the *Quick Select* feature to identify all circles with a radius value over 20 on the drawing.

1.   Press the Quick Select button to activate the *Quick Select* dialogue box.

The *Object type:* edit box indicates that multiple objects are present in the drawing.

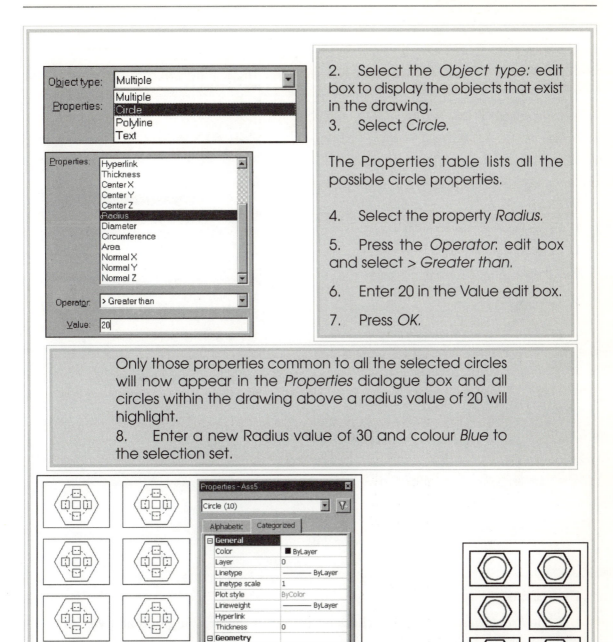

2. Select the *Object type:* edit box to display the objects that exist in the drawing.

3. Select *Circle*.

The Properties table lists all the possible circle properties.

4. Select the property *Radius*.

5. Press the *Operator:* edit box and select > *Greater than*.

6. Enter 20 in the Value edit box.

7. Press *OK*.

Only those properties common to all the selected circles will now appear in the *Properties* dialogue box and all circles within the drawing above a radius value of 20 will highlight.

8. Enter a new Radius value of 30 and colour *Blue* to the selection set.

Now you will create a new selection set.

9. Select the *Apply to:* edit box and then select *Entire drawing*.

10. Make the *Object type: Polyline*.

11. Select the *Property: Area*.

12. Set the *Operator to >Greater than*.

13. Enter a *Value* of 5000.

14. Press *OK*.

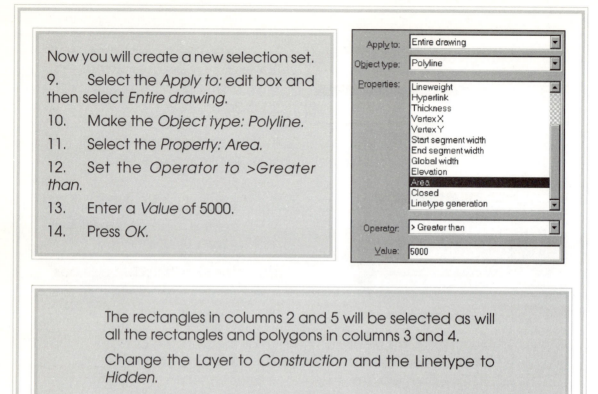

The rectangles in columns 2 and 5 will be selected as will all the rectangles and polygons in columns 3 and 4.

Change the Layer to *Construction* and the Linetype to *Hidden*.

Press the *ESC* key twice to clear the selection set and close the *Properties* dialogue box.

Exercise

Use the *Properties* command to change the text in the drawing to text style *ISO* with a text height of 20.

Chapter 14

Grips

The object of this chapter is to illustrate how you can edit selected objects by manipulating Grips that appear at defining points on the objects.

A Grip is a small square that appears at various object-specific positions, for instance, at the quadrants and centre of a circle or endpoints of a line if an object is selected outside of a command. Gripped objects can be edited using the following commands: Stretch, Move, Rotate, Scale and Mirror.

Objectives

At the end of this chapter you will be able to:

▷   Understand the concept of the Grips feature.

▷   Edit selected objects using the Grips feature.

▷   Enable or disable Grips using the Options dialogue box.

New command

▷   DDGrips   GR

The Grips feature

The Grips feature enables you to edit selected objects by manipulating defining points on the object. A Grip is a small square that appears at various object-specific positions such as circle centres or quadrants and the endpoint and midpoint of a line.

The following diagram illustrates the location of Grips in a selection of objects.

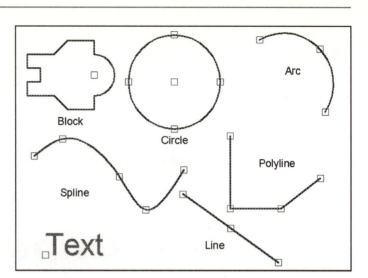

Activating grips

To activate grips, select the object with the graphics cursor when no command is active i.e. when the prompt displays Command.

The graphics cursor will automatically snap to a grip when it moves over one. Select any one of the grip boxes to make it *Hot*. This will turn the grip red in colour and enable the various grip options.

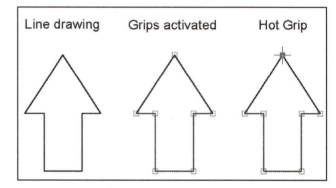

Click the right mouse button to activate the Grip cursor menu that will allow you to select a variety of editing options.

When a grip is hot (red) you can also cycle through the commands by pressing the space bar. To clear grips, press *ESC* key twice to cancel the command.

Editing with grips

The Grips feature allows you to Stretch, Move, Rotate, Scale and Mirror selected objects.

Example – stretching

1.   With no command active select the Polyline object to activate its Grips.
2.   Select the grip at the point to be stretched to make it *hot*.
3.   Right click the pointing device to activate the *Grips* menu.
4.   Select the *Stretch* option.
5.   Move the hot grip to the required location.

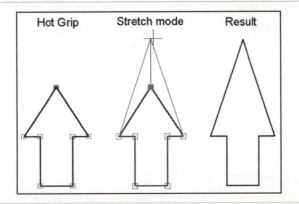

Hot Grip   Stretch mode   Result

Example – move and copy

The following example will illustrate how *Grips* move a copied version of the selected object.

1.   With no command active select the *Polyline* object to activate its *Grips*.
2.   Select the grip at the point to be moved to make it *hot*.
3.   Right click the pointing device to activate the *Grips* menu and select the *Move* option.
4.   Right click the pointing device again to activate the *Grips* menu and select the *Copy* option.
5.   Move the hot grip to the required location then left click the pointing device.

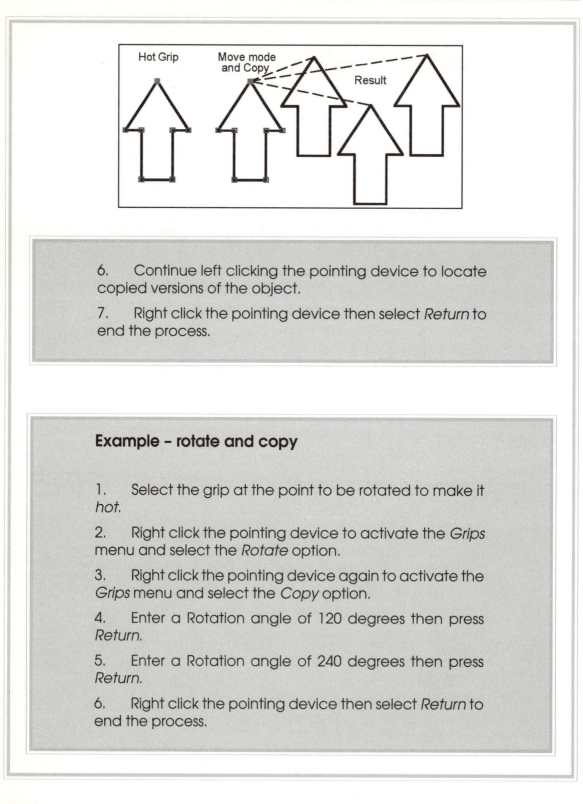

6. Continue left clicking the pointing device to locate copied versions of the object.

7. Right click the pointing device then select *Return* to end the process.

Example – rotate and copy

1. Select the grip at the point to be rotated to make it *hot.*

2. Right click the pointing device to activate the *Grips* menu and select the *Rotate* option.

3. Right click the pointing device again to activate the *Grips* menu and select the *Copy* option.

4. Enter a Rotation angle of 120 degrees then press *Return.*

5. Enter a Rotation angle of 240 degrees then press *Return.*

6. Right click the pointing device then select *Return* to end the process.

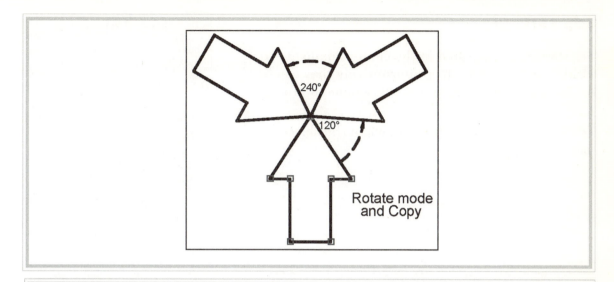

Rotate mode
and Copy

Exercise

Create and modify a simple object using the Mirror and Scale Grip options.

Fine tuning using grips

Because grips appear on object features such as endpoints and midpoints, snapping to grips is an alternative to using object snaps for editing commands.

Example 1

Using the Grip Move option move the *hot* grip to snap to another line endpoint grip.

Example 2

Use the grips located on a dimension block to fine tune the position of either the dimension text, the extension lines or both.

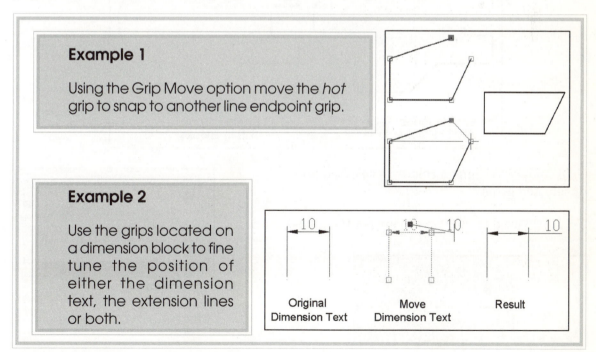

Original
Dimension Text

Move
Dimension Text

Result

Grip settings

Grip settings are located in the *Options* dialogue box within the *Selection* tab or alternatively they can be accessed by typing *Ddgrips* at the command prompt.

Enable grips: grips will activate when the object is selected.

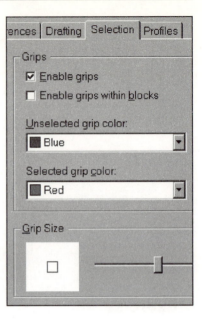

Enable grips within blocks: turned *on* it will assign grips to objects within a block. If turned *off*, the block is assigned one grip.

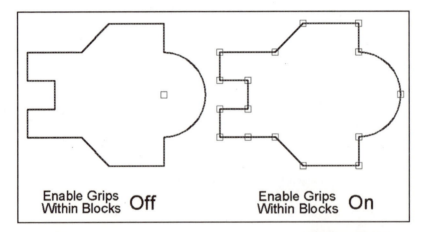

Enable Grips Within Blocks **Off**   Enable Grips Within Blocks **On**

Grip colours: assigns a colour to selected and unselected grips.

Grip size: used to change the size of the grips.

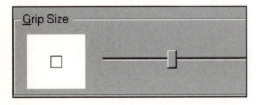

Chapter 15

Blocks

A Block is a series of objects (lines, arcs, text etc.) grouped together to form a single entity. This entity can be inserted into a drawing any number of times without significantly increasing the size of the drawing file. You can use Blocks to create libraries of frequently used symbols such as mechanical fasteners, electronic symbols, architectural doors and windows etc.

The following chapter illustrates the advantages of using Blocks.

Objectives

At the end of this section the user will be able to:

▷   Understand the concept and use of the Blocks and Wblocks.

▷   Create and insert blocks into a drawing.

▷   Modify a Block definition.

New commands

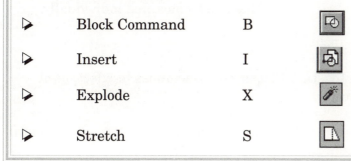

▷	Block Command	B	
▷	Insert	I	
▷	Explode	X	
▷	Stretch	S	

Blocks

Draw toolbar	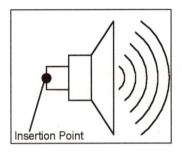
Draw menu	**Block**
Command line entry	**Block**
Alias	**B**

A Block definition is a collection of objects grouped together to form a single object that can be inserted into a drawing any number of times. When inserted, blocks can be assigned different scale and rotation factors.

A typical example of a block could be the speaker drawing shown.

This speaker can be inserted into the drawing as many times as is necessary.

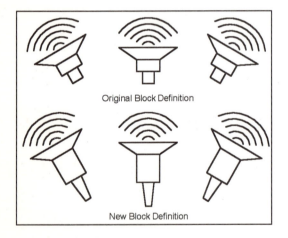

Original Block Definition

New Block Definition

Blocks versus the Copy command

Revising drawings can be time consuming if you have to locate and edit several copies of an affected part. However, if the part is defined as a Block, it can simply be exploded and redefined. All references to the Block will be updated automatically.

In other words, if the speaker needed to be modified then changes to its Block definition will apply to all the other speakers in the drawing. Had the original speaker simply been copied, then each copy would have to be modified individually.

Further advantages of using blocks include:

Drawings can be created by adding together Blocks, sub assemblies in effect, used together to create larger components.

Blocks can be used to build up a library of frequently used symbols, blocks or standard drawings i.e. the mechanical fasteners shown.

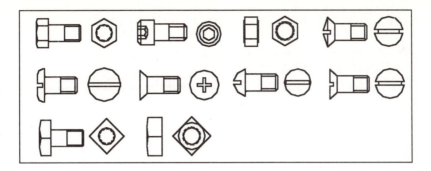

Each object added to a drawing increases the size of the drawing file on the disk. Multiple instances of a Block are stored as one reference in the drawing database, thereby reducing file size.

In the age of the Internet and the communication of drawing across large distances, drawing size becomes ever more important.

Block definitions can be loaded into any number of drawings irrespective of where they were created. You can use AutoCAD DesignCenter (see Chapter 18) to browse and locate the block you want to copy.

Block command

The Block command is used to create new Blocks that will be defined within the current drawing file.

From the *Draw* menu select *Block* then *Make...*

This will activate the *Block Definition* dialogue box:

Name – Specifies the name of the block up to 31 characters.

Base Point – Indicates the insertion point for the block. Specify X,Y,Z co-ordinates or use the pick point button to specify base point on the drawing.

Select Objects – Picks the objects to include in the new block.

Retain – If checked, selected objects will remain in the drawing.

Convert to block – Converts and replaces the selected objects to a block.

Delete – If checked objects included in the block will be deleted from the drawing.

Preview icon – View preview icon On or Off.

Units – Specifies the units the block is scaled to.

Description – Accepts a textual description up to 256 ASCII characters long for display in Content Explorer. This description can be modified with *Bmod*.

Example - creating a block

1.   Start a *New* drawing using the *Student – Template* option.
2.   Create the following drawings. Do not include the dimensions.

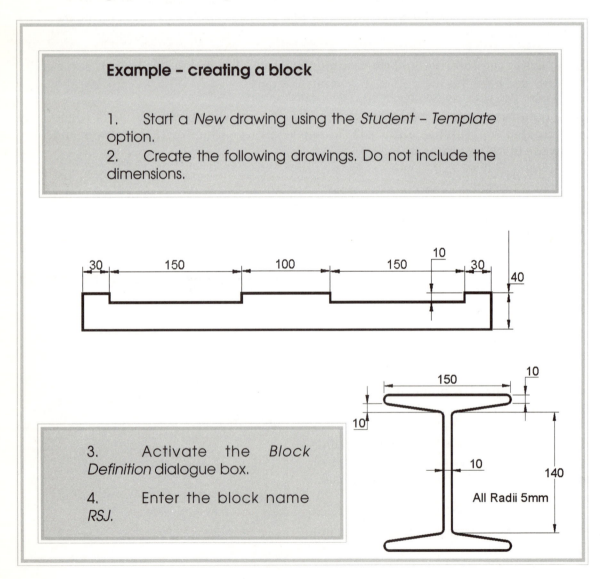

3.   Activate the *Block Definition* dialogue box.

4.   Enter the block name *RSJ*.

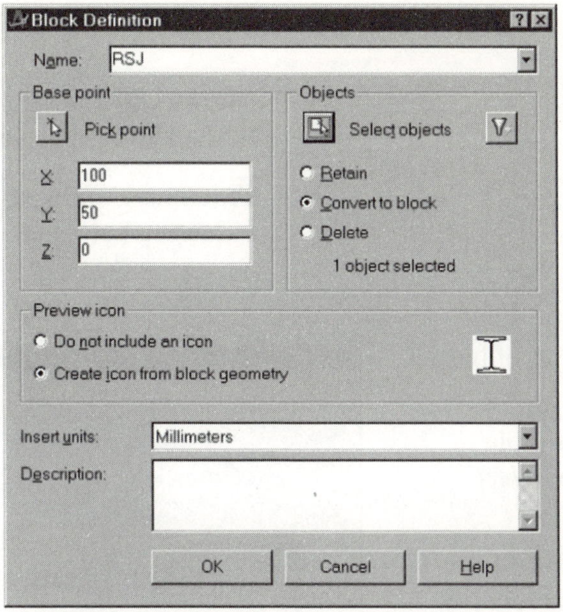

Block Definition

Name: RSJ

Base point

Pick point

X: 100

Y: 50

Z: 0

Objects

Select objects

○ Retain
● Convert to block
○ Delete

1 object selected

Preview icon

○ Do not include an icon
● Create icon from block geometry

Insert units: Millimeters

Description:

[OK] [Cancel] [Help]

---

5.   Use the *Pick point* button and pick the midpoint of the base of the *RSJ*.

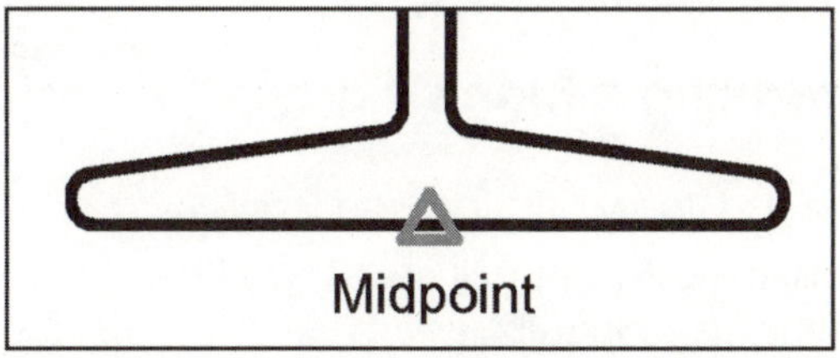

Midpoint

---

6.   Use the *Select objects* button to select the *RSJ*.

7.   Check the *Delete* option.

8.   Press *OK* to exit.

○ Delete

Block insertion

Draw toolbar	🔲
Insert menu	**Block...**
Command line entry	**Insert**
Alias	**I**

Blocks can be inserted into a drawing by using the *Insert* dialogue box.

The *Insert* dialogue box includes the following settings:

Name – Select from a list of blocks.

Browse – Choose Browse... to select from a list of available drawing files using the *Select Drawing File* dialogue box.

Specify on-screen – Select to use the pointing device to define the *Insertion point*, *Scale* and *Rotation* angle of the block on the drawing. Clear to set the insertion point, scale, and rotation angle with the options that follow.

Explode – Select to insert the block as the individual objects that make up the block.

Example – inserting a block

1. Activate the *Insert* dialogue box.
2. The Name: edit box will show *RSJ* because it is the defined block in the drawing.
3. Ensure *Insertion point* is set to *Specify on-screen* then press *OK*.

4.   Specify insertion point: locate the *RSJ* at the midpoint of the lower left slot.

5.   Repeat and locate the second *RSJ* at the midpoint of the lower right slot.

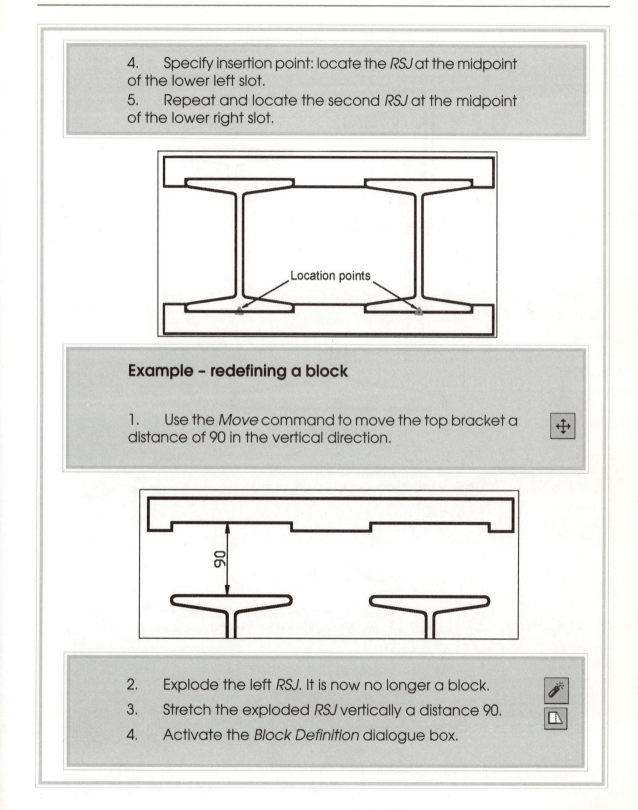

Location points

Example – redefining a block

1.   Use the *Move* command to move the top bracket a distance of 90 in the vertical direction.

2.   Explode the left *RSJ*. It is now no longer a block.

3.   Stretch the exploded *RSJ* vertically a distance 90.

4.   Activate the *Block Definition* dialogue box.

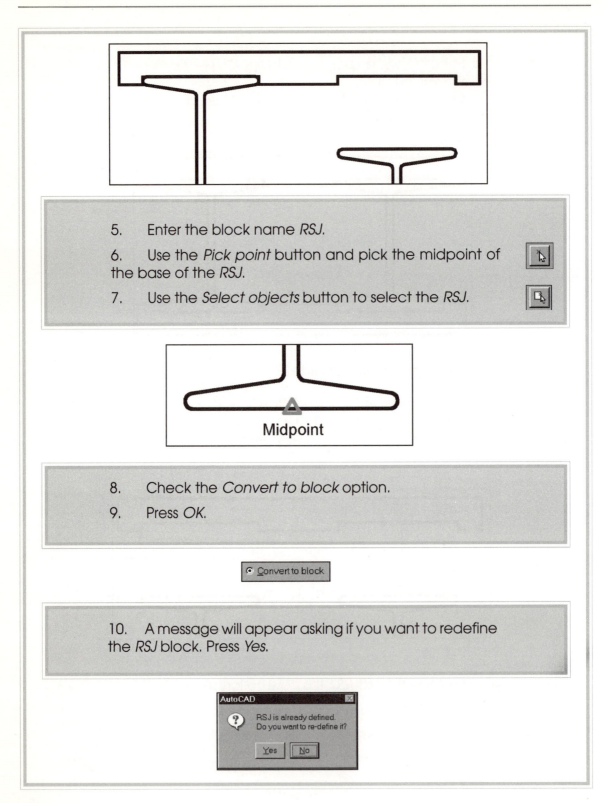

5. Enter the block name *RSJ*.

6. Use the *Pick point* button and pick the midpoint of the base of the *RSJ*.

7. Use the *Select objects* button to select the *RSJ*.

Midpoint

8. Check the *Convert to block* option.

9. Press *OK*.

◉ Convert to block

10. A message will appear asking if you want to redefine the *RSJ* block. Press *Yes*.

AutoCAD

RSJ is already defined.
Do you want to re-define it?

Yes   No

11.   The right hand *RSJ* will update to the new block definition.

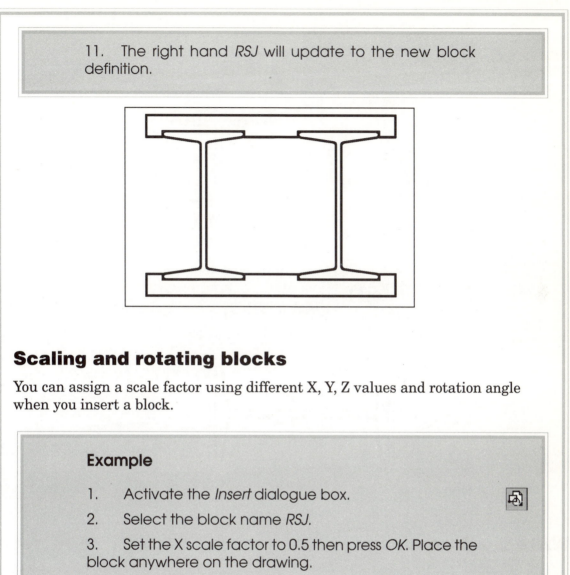

Scaling and rotating blocks

You can assign a scale factor using different X, Y, Z values and rotation angle when you insert a block.

Example

1.   Activate the *Insert* dialogue box.

2.   Select the block name *RSJ*.

3.   Set the X scale factor to 0.5 then press *OK*. Place the block anywhere on the drawing.

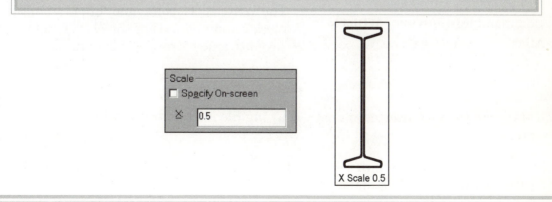

X Scale 0.5

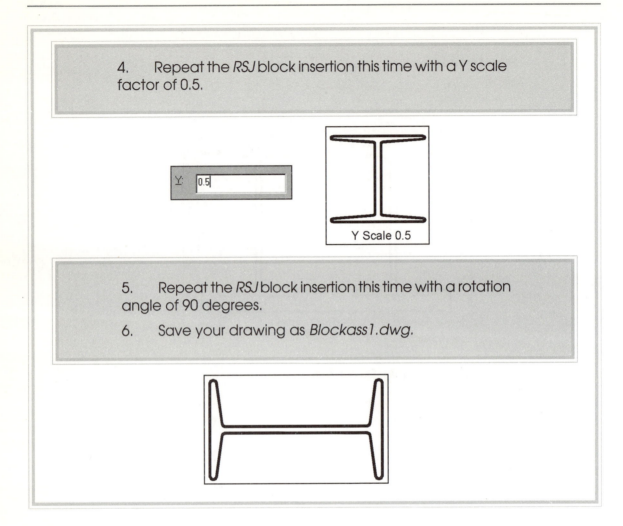

4. Repeat the *RSJ* block insertion this time with a Y scale factor of 0.5.

Y: 0.5

Y Scale 0.5

5. Repeat the *RSJ* block insertion this time with a rotation angle of 90 degrees.

6. Save your drawing as *Blockass1.dwg*.

Write block command

Command line entry	WBlock
Alias	W

The Write Block command differs from the block command in that the block created is saved to a new drawing file in its own right.

The block can then be inserted into any drawing file irrespective of where the block was created.

You can turn your drawing into a *Wblock* by typing *Wblock* at the command line. This will activate the *Write Block* dialogue box.

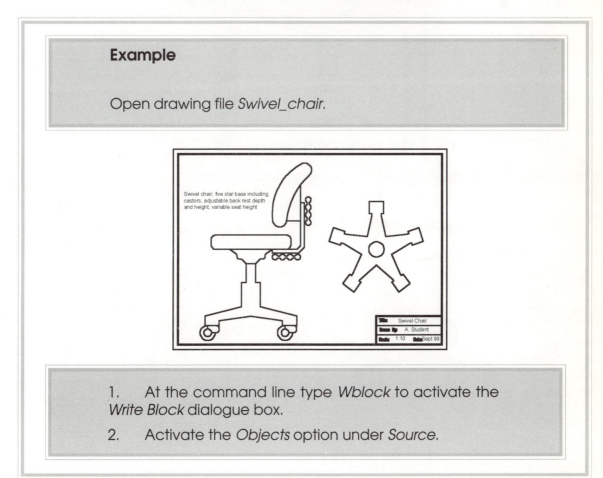

Example

Open drawing file *Swivel_chair*.

Swivel chair, five star base including
castors, adjustable back rest depth
and height, variable seat height

Title	Swivel Chair
Drawn By	A. Student
Scale	1:10	Date	Sept 99

1. At the command line type *Wblock* to activate the *Write Block* dialogue box.

2. Activate the *Objects* option under *Source*.

3. Using the *Pick point* button specify the base point as the centre of the left castor.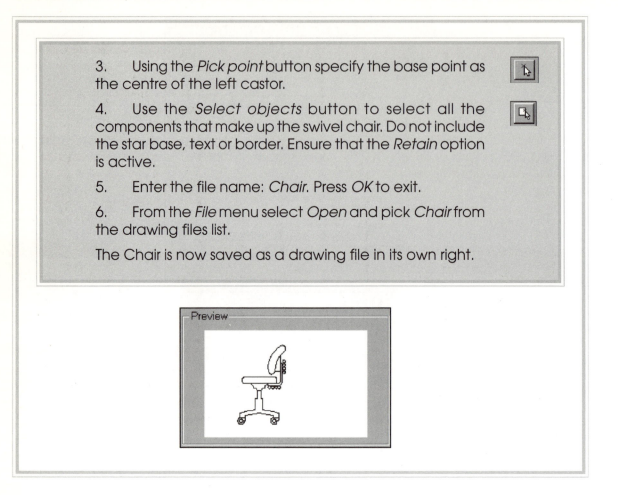

4. Use the *Select objects* button to select all the components that make up the swivel chair. Do not include the star base, text or border. Ensure that the *Retain* option is active.

5. Enter the file name: *Chair*. Press *OK* to exit.

6. From the *File* menu select *Open* and pick *Chair* from the drawing files list.

The Chair is now saved as a drawing file in its own right.

Preview

Chapter 16

Attributes

An Attribute provides an interactive label or tag that allows you to attach text or other data to a Block. Whenever you insert a Block that has an Attribute attached to it, you are prompted to enter the data that will be stored with the Block. Examples of data may be parts of names, prices and catalogue numbers etc. The following chapter will take you step by step through the attribute creation process and concludes with a practical exercise designed to test your understanding of the features covered.

Objectives

At the end of this exercise you will be able to:

▷   Understand what is meant by an Attribute.

▷   Use the Attribute dialogue box.

▷   Create a block with attributes.

▷   Complete a practical exercise using attributes.

New command

Attribute Definition   ATTDEF

Attribute definition

Draw toolbar	🏷️
Draw menu	**Block ▶**
	Define attributes
Command line entry	**Attdef**
Alias	**ATT**

An attribute is informational text associated with a block. An attribute definition specifies an attribute's properties and the prompts which appear when the block is inserted.

The *Attribute Definition* dialogue box creates the attribute tag that appears in the drawing. When a block is inserted, the attribute tag is replaced by the attribute value at the same location in the block with the same text style and alignment.

Activate the *Attribute Definition* dialogue box to create an attribute definition, which describes the characteristics of the attribute. The characteristics include the tag, prompt, value information, text formatting and location.

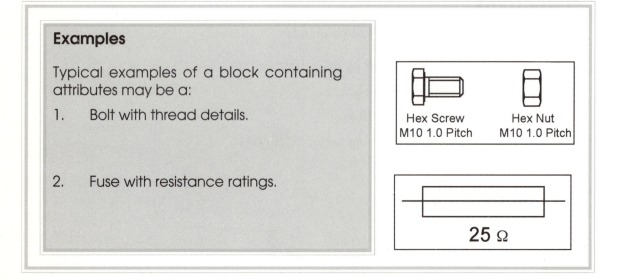

Examples

Typical examples of a block containing attributes may be a:

1. Bolt with thread details.

2. Fuse with resistance ratings.

Hex Screw
M10 1.0 Pitch

Hex Nut
M10 1.0 Pitch

25 Ω

3.   Telephone with name and extension details.

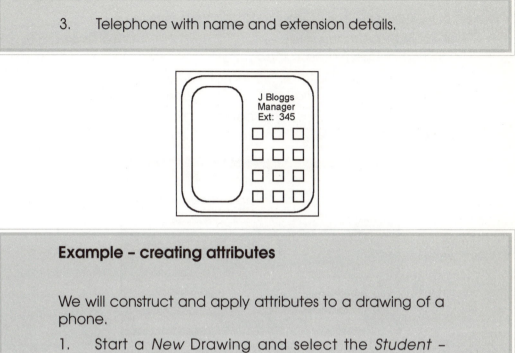

Example – creating attributes

We will construct and apply attributes to a drawing of a phone.

1.   Start a *New* Drawing and select the *Student – Template* option.

2.   Create the drawing of a telephone as illustrated. Do not dimension the drawing.

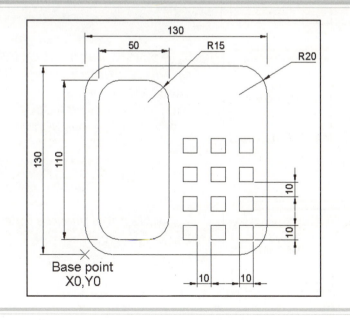

3. Activate the *Attribute Definition* dialogue box.

4. In the *Attribute* area enter *Name* in the *Tag:* edit box and *Enter User Name* in the *Prompt:* edit box.

5. In the Text Options area make text *Justification – Middle, Text style – Arial,* and text *Height* 5.

6. Press the *Pick point* < button and pick the position at co-ordinates 95,110.

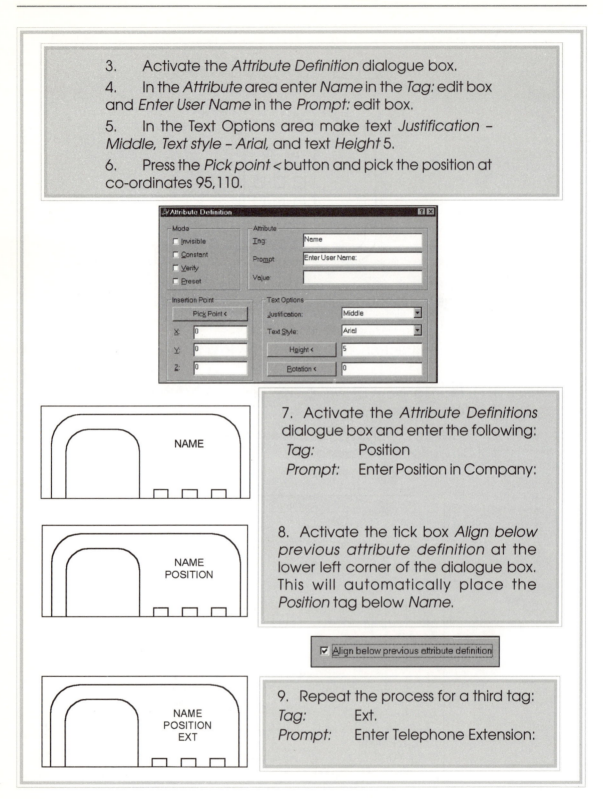

7. Activate the *Attribute Definitions* dialogue box and enter the following:
Tag: Position
Prompt: Enter Position in Company:

8. Activate the tick box *Align below previous attribute definition* at the lower left corner of the dialogue box. This will automatically place the *Position* tag below *Name*.

9. Repeat the process for a third tag:
Tag: Ext.
Prompt: Enter Telephone Extension:

Your next task is to turn the telephone and the attributes into a Block.

Example – creating a block with attributes

1. Activate the *Block Definition* dialogue box by entering *B* at the command prompt.

2. In the *Name:* edit box enter *Phone*.

Name: Phone

3. Enter the following co-ordinates in the Base point edit boxes:

X: 65

Y: 65

4. Press the *Select objects* button and pick all of the rectangles that make up the telephone then pick the attribute tags in the order they were created.

Base point

Pick point

X: 65

Y: 65

Z: 0

Objects

Select objects

○ Retain

○ Convert to block

● Delete

17 objects selected

Preview icon

○ Do not include an icon

● Create icon from block geometry

Note – placing a selection window around the whole telephone would result in the tag prompts appearing in reverse order i.e. the last tag created would be the first to be prompted.

5. Ensure that the *Delete* option is active in the *Objects* area of the dialogue box then press *OK*. The telephone and attributes will disappear from the drawing.

Example – inserting a block with attributes

1. At the command prompt enter *Attdia* and set the variable to 1.

Note – this will enable the Enter Attribute dialogue box to appear on the drawing area. Attdia set to O would result in the attribute prompts appearing at the command line when the block is inserted.

2. From the *Insert* menu select *Block...*
3. Select *Phone* from the *Name:* edit box then press *OK*.

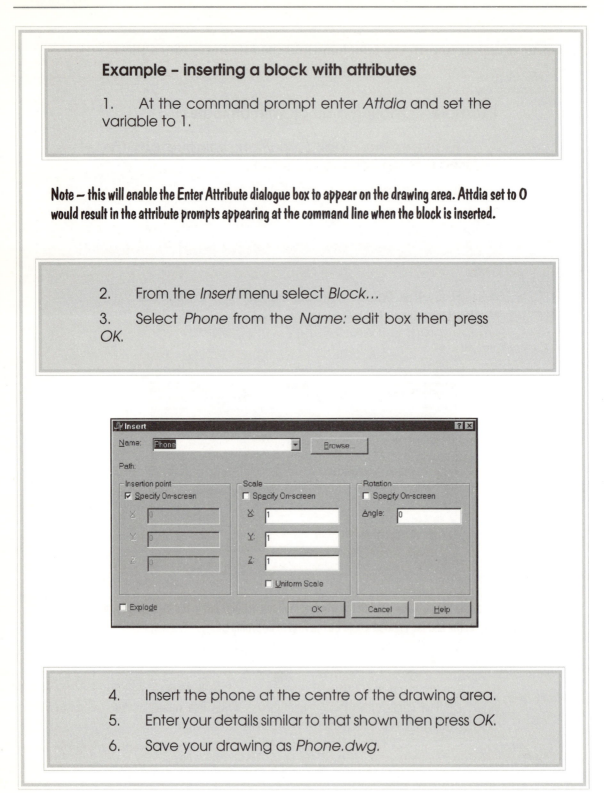

4. Insert the phone at the centre of the drawing area.
5. Enter your details similar to that shown then press *OK*.
6. Save your drawing as *Phone.dwg*.

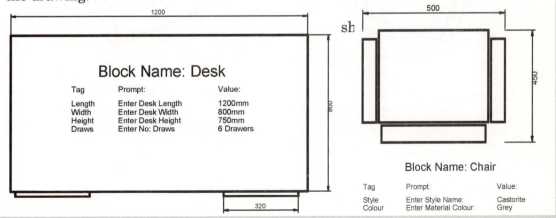

Build upon the current drawing by completing the blocks and attributes in the following exercise.

Exercise – using blocks and attributes

Create the following drawings. Use your discretion concerning dimensions. Convert them to blocks containing the attributes indicated. In each case, select a suitable text style and height and position the attributes within the extents of the drawing.

Block Name: Desk

Tag	Prompt:	Value:
Length	Enter Desk Length	1200mm
Width	Enter Desk Width	800mm
Height	Enter Desk Height	750mm
Draws	Enter No: Draws	6 Drawers

Block Name: Chair

Tag	Prompt:	Value:
Style	Enter Style Name:	Castorite
Colour	Enter Material Colour:	Grey

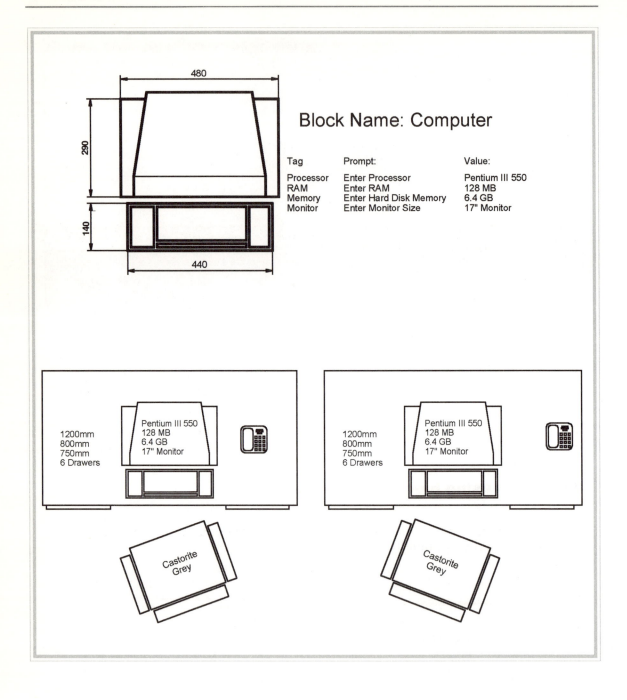

Block Name: Computer

Tag	Prompt:	Value:
Processor	Enter Processor	Pentium III 550
RAM	Enter RAM	128 MB
Memory	Enter Hard Disk Memory	6.4 GB
Monitor	Enter Monitor Size	17" Monitor

Chapter 17

Model space and Paper space

The following chapter will introduce you to the concept of Model space and Paper space view modes. Model space is the environment in which drawings are created. Paper space provides viewports from which the drawing may be viewed.

A viewport is used to control the viewing area of the drawing, scale factor and layer visibility. Viewports are created within Layouts which represent the size and orientation of the paper onto which the drawing is printed.

In the course of this chapter, you will create viewports within a number of Layouts and, using the Print feature, produce printed copies of your drawing.

Objectives

At the end of this section the user will be able to:

- ▷ Understand the difference between Paper space and Model space modes.
- ▷ Select Multiple viewport configurations.
- ▷ Alter the Zoom factor within viewports.
- ▷ Create new Layouts.
- ▷ Convert closed polyline objects into viewports.
- ▷ Manipulate the position and size of these viewports.
- ▷ Use the Print feature to produce printed copies of your work.

New commands

- ▷ Zoom Extents
- ▷ Display Viewports
- ▷ Polygonal Viewport
- ▷ Convert Object to Viewport
- ▷ Print

Model space and Paper space

When you create drawings you can operate in either Model space or Paper space.

You use Model space (the Model tab) to create two-dimensional drawings or three-dimensional models and Paper space (a layout tab) to create a finished layout of a drawing for plotting.

Think of Paper space as representing the paper in your plotter or printer onto which you place a number of viewports that contain various views of the drawing. You control the number, size, scaling factor and layer visibility of each viewport.

The illustrations below show a completed drawing in Model space first and a layout of various views of the drawing in Paper space second.

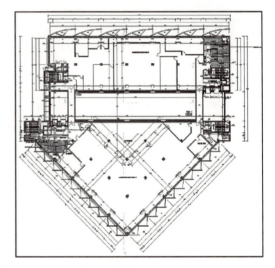

Model space

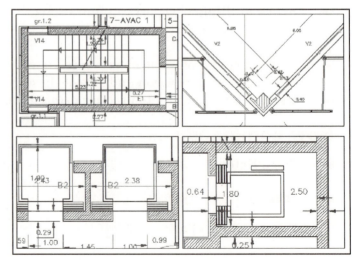

Paper space viewports

To illustrate the concept of Model space and Paper space you will use an existing drawing from the *Acad2000/Sample* folder.

Example

1.   Open drawing file *db_samp.dwg* from the *Acad2000\Sample* folder.
2.   Use the *Zoom Extents* tool in the *Zoom* dropdown toolbar to make the drawing fill the screen.
3.   The drawing is currently in Model space. Your next task is to create and layout 4 viewports in Paper space. Each viewport will present a specific area of the drawing.

4. To enter Paper space, press the *Layout1* tab at the base of the graphics area.

5. By default, AutoCAD has created a single viewport that represents the drawing in Model space.

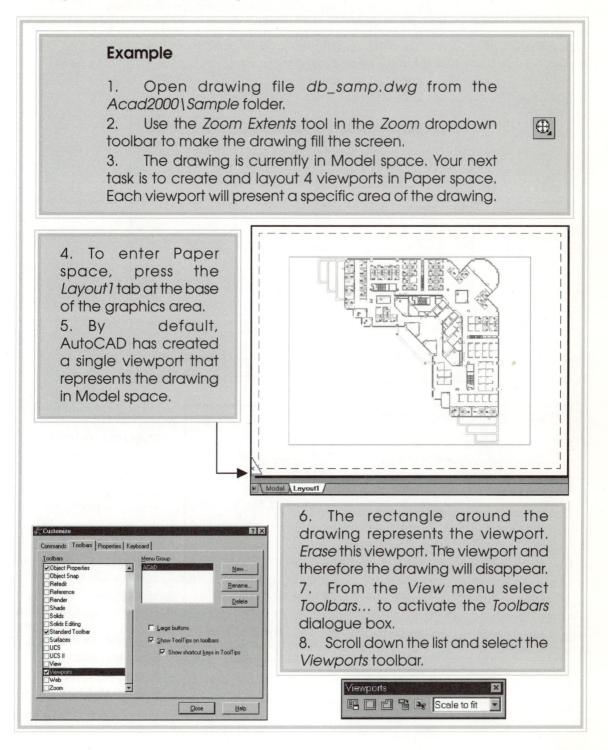

6. The rectangle around the drawing represents the viewport. *Erase* this viewport. The viewport and therefore the drawing will disappear.

7. From the *View* menu select *Toolbars...* to activate the *Toolbars* dialogue box.

8.   Scroll down the list and select the *Viewports* toolbar.

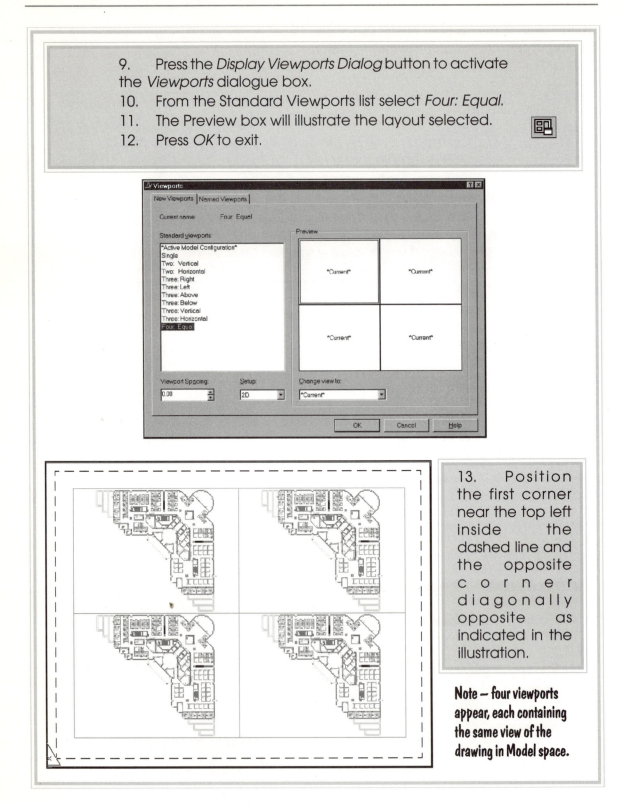

9. Press the *Display Viewports Dialog* button to activate the *Viewports* dialogue box.
10. From the Standard Viewports list select *Four: Equal*.
11. The Preview box will illustrate the layout selected.
12. Press *OK* to exit.

13. Position the first corner near the top left inside the dashed line and the opposite corner diagonally opposite as indicated in the illustration.

Note – four viewports appear, each containing the same view of the drawing in Model space.

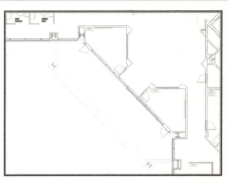

14. You can enter any selected viewport within the layout, to work in Model space. To switch to Model space double-click a viewport. You can switch back to Paper space by double-clicking an area of Paper space.

15. Double click anywhere within the top left viewport. The rectangular border will highlight.

16. Use Zoom window to select an area of the drawing you want to fill the viewport.

17. Single click into the three remaining viewports in each case zooming a window around a selected area of your choice.

Note – you can also switch between floating Model space view and Paper space by choosing Model or Paper on the status bar.

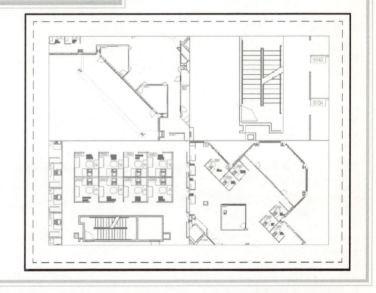

Example – creating a new layout

The next layout will consist of three viewports which will be created using the various *Viewports* toolbar options.

1.   Right click on the *Layout1* tab and select *New Layout*. A new tab *Layout2* will appear.

2.   Select the tab *Layout2*. This will activate the *Page Setup* dialogue box. The *Page Setup* dialogue box has two tabs:

Plot device
Used to name and save the current page setup and specify your currently configured plotting device to which you will send a layout for plotting.

Layout settings
Used to specify layout settings such as:
● Plot area
● Plot scale
● Plot offset
● Drawing orientation
● Paper size.

3.	Accept all the defaults and press *OK* to exit *the Page Setup* dialogue box.
4.	*Erase* the single viewport that defaults within the layout.
5.	Select the *Polygonal Viewport* tool from the *Viewports* toolbar.

Note – the Polygonal Viewport tool allows you to create a viewport using a closed polyline.

6.	Draw a polygonal viewport similar to that illustrated on the left of the layout. Do not be concerned with dimensional accuracy.

Note – the Convert Object to Viewport tool will convert any existing closed polyline into a viewport.

7.	Create a circle in the layout as indicated.
8.	Press the *Convert Object to Viewport* button and select the circle.

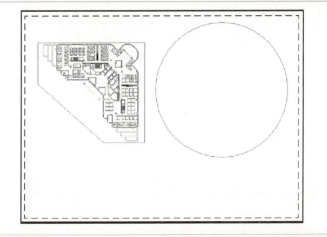

9.	The circle will become a viewport. Use your discretion to zoom and pan a required view.
10.	Create a Spline similar to that illustrated below. Use the *Convert Object to Viewport* button to change the spline into a viewport. Zoom and pan as required.

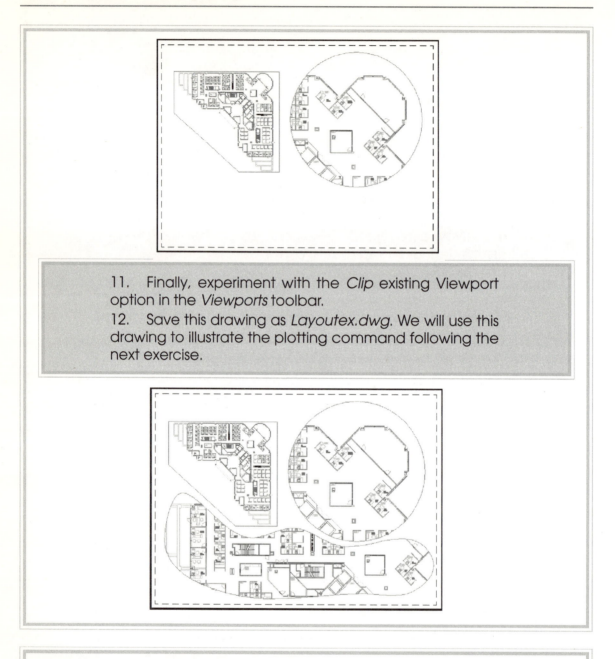

11. Finally, experiment with the *Clip* existing Viewport option in the *Viewports* toolbar.

12. Save this drawing as *Layoutex.dwg*. We will use this drawing to illustrate the plotting command following the next exercise.

Exercise – creating new layouts

Open the drawing file *Oceanarium.dwg* from the *Acad2000/Sample* folder.

Create a new layout, paper size A4, Landscape. Produce the viewports illustrated in Paper space.

Use your own discretion concerning the views presented within the viewports.

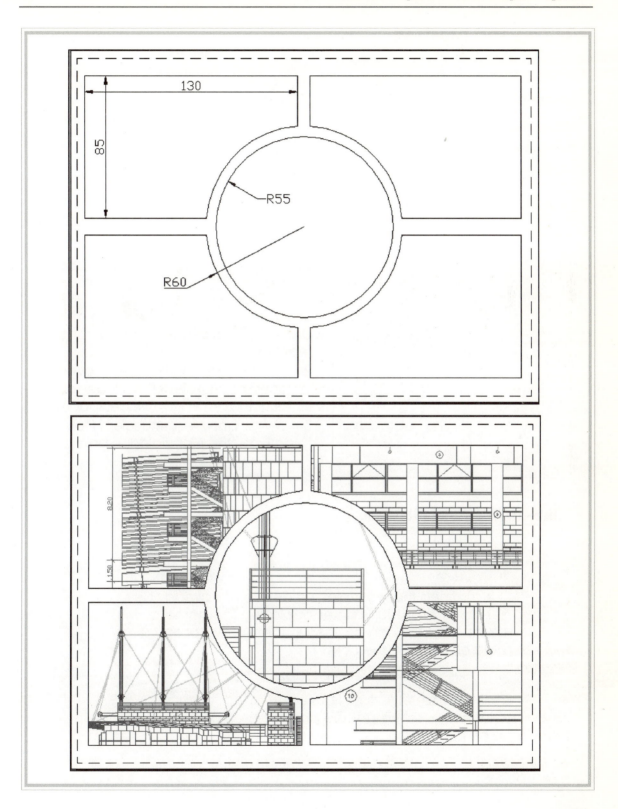

Printing and plotting

Standard toolbar	🖨
File menu	**Plot...**
Command line entry	**Plot**

The Plot command is used to plot a drawing to a printer or plotting device or save the drawing as a plot file.

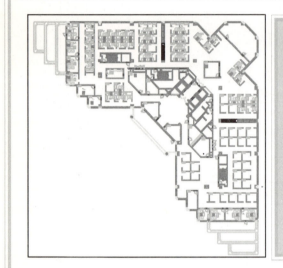

Example – plotting in Model space

1. Open drawing file *Layoutex1.dwg*. We will use this drawing to illustrate the plotting features.

2. If the drawing has been saved in Paper space, select the *Model* tab to return the drawing to Model space.

3. From the *File* menu select *Plot...* to activate the *Plot* dialogue box.

Note – the *Plot Device* tab will be current.

It is assumed that your plotting device has already been set up to a default printer or plotter. The name of the plotting device can be seen in the *Plotter configuration* area of the dialogue box.

4. Select the *Plot Settings* tab.

The *Plot Settings* tab is used to specify paper size, orientation, drawing scale and the area of the drawing to be plotted.

5.   Set your paper size to *A4 210,297mm* and orientation to *Landscape*.

6.   Select the *Plot scale* option *Scaled to fit.* This will scale the area of the drawing that is selected to fit the plotted paper.

There are four Plot area options. We will look at them in turn.

Limits: Will plot only that part of the drawing that exists within the drawing limits. Large drawings can exceed the limits of the drawing and objects outside the drawing limits will not print.

7.   Press the *Full Preview* button to see the area of the drawing that will be printed.
All the objects within the drawing are currently outside the drawing limits therefore no part of the drawing will print.

> 8.   Right click to activate the cursor menu and pick *Exit* to return to the *Plot* dialogue box.

Extents: Will plot the extents of the drawing.

> 9.   Press the *Full Preview* button to see the area of the drawing that will be printed.

Display: Will plot the view in the current viewport.

> 10.   Press the *Full Preview* button to see the area of the drawing that will be printed.

View: Will plot a view saved previously with the *View* command.

Window<:   Will plot any portion of the drawing you specify by two corners.

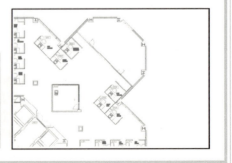

> 11.   Press the *Window<* button. You will be prompted to specify the area of the drawing required for plotting by two corners. Select the top right corner of the drawing.

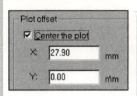

12. You will notice that the selected area of the drawing is located at the left side of the paper. You can centre the drawing plot by activating the *Center the plot* tick box in *the Plot offset* area of the dialogue box.

13. When satisfied with your selection press *OK* to print.

Example – plotting in Paper space

1.   Press the *Layout1* tab. Four viewports will appear each with a different view of the drawing.

2.   Activate the plot command. Ensure that the *Plot Settings* tab is active.

3.   You will notice in the *Plot area* part of the tab that the *Layout* button is active. Press the *Full Preview* button to view the area of the drawing to be plotted.

You will see that the layout fits the available plotting area.

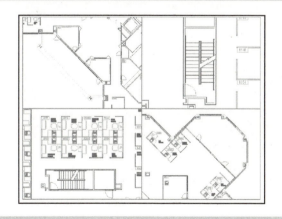

4.   Press the *Partial Preview* button. This will highlight the plot area relative to the paper size and printable area.

5.   Press *OK* to print the layout.

6.   Activate the *Layout2* tab and print the complete layout.

7.   *Save* and exit your drawing.

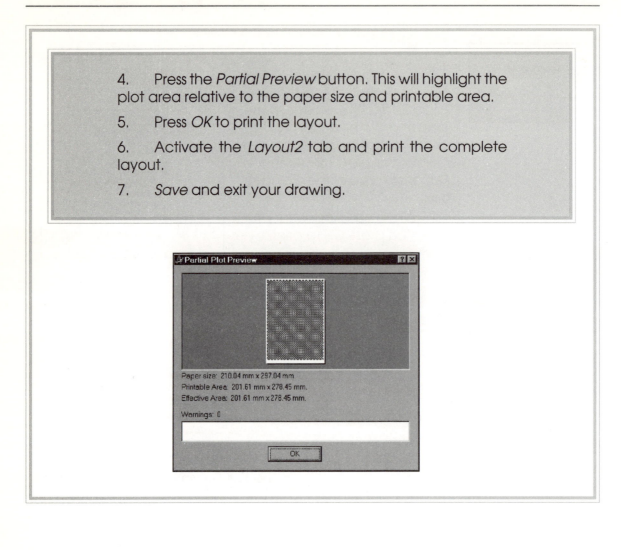

Chapter 18

AutoCAD DesignCenter

The AutoCAD DesignCenter is used to locate and insert into your current drawing, objects that exist in other drawing files. Objects that can be inserted include blocks, dimension styles, text styles, layers, linetypes and xrefs.

Drawing files containing libraries of blocks such as mechanical fasteners or electronic symbols, can be accessed immediately which enhance drawing productivity and standardisation. The following chapter will introduce you to the benefits of the DesignCenter by taking you through the various steps of inserting objects from external file sources into your current drawing.

Objectives:

At the end of this section you will be able to:

▷   Understand the operating principles of the DesignCenter.

▷   Insert objects from external drawing files into your current drawing.

New commands

▷   AutoCAD DesignCenter   ADC   ▷   ADC Load

▷   ADC Desktop   ▷   ADC Find

▷   ADC Open drawing   ▷   ADC Preview

AutoCAD DesignCenter

Standard toolbar	
Tools menu	**AutoCAD DesignCenter**
Command line entry	**ADCENTER**
Alias	**ADC**

The DesignCenter is used to browse and locate objects in a drawing such as blocks, xrefs, layouts, layers, dimstyles and textstyles that can be previewed and inserted into the current drawing irrespective of their original location.

The DesignCenter, when active, is a separate window which by default is docked on the left side of the graphics screen. It can be floated into any position as necessary.

It consists of a tree view of folders and drawings and a palette of large icons.

The palette display is dependant upon the item selected in the tree view.

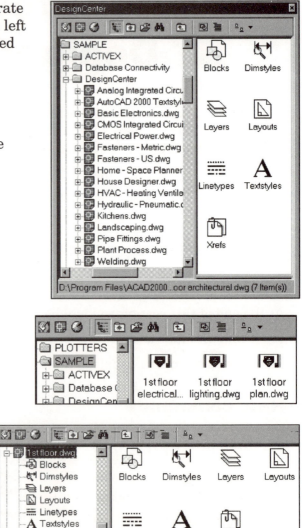

1. Selecting a folder in the tree view will result in the image icons of the drawings within that folder appearing in the palette.

2. Selection of the + sign at the left of each drawing file will display a list of the objects located within the drawing.

3. Clicking on *Blocks* in either the tree view or the palette will reveal an image icon of the blocks defined within the drawing.

4. Click on *Dimstyles* in either the tree view or palette to reveal the different dimension styles defined within the drawing.

5. *Layers* display the layer names within the drawing.

6. *Layouts* display layout names.

7. *Linetypes* display loaded linetypes in the drawing.

8. *Textstyles* display the name of textstyles in the drawing.

9. *Xrefs* display the Xrefs attached to the drawing.

DesignCenter toolbar options

Desktop – Displays the files and folders on your computer.

Open Drawings – Displays all drawings currently open in the AutoCAD session, including drawings that are minimised.

History – Displays up to the last 20 Web locations to which you connected.

Tree View Toggle – Displays and hides the tree view.

Favorites – Displays the contents of the *Favorites* folder in the palette.

Load – Displays the *Load DesignCenter* Palette dialogue box. Objects within a selected file will appear in the palette.

Find– Displays the *Find* dialogue box, in which you can specify search criteria to locate objects within drawings.

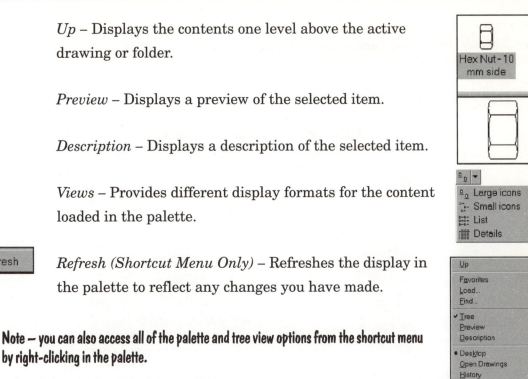

Up – Displays the contents one level above the active drawing or folder.

Preview – Displays a preview of the selected item.

Description – Displays a description of the selected item.

Views – Provides different display formats for the content loaded in the palette.

Refresh (Shortcut Menu Only) – Refreshes the display in the palette to reflect any changes you have made.

Note – you can also access all of the palette and tree view options from the shortcut menu by right-clicking in the palette.

We can see that the DesignCenter allows you to set up and create new drawings by taking existing objects from other drawings. You can set up your new drawing, in effect like a jigsaw puzzle, by inserting bits and pieces from various files.

The following example will illustrate how different types of objects are inserted into the current drawing. Using the DesignCenter you will insert:

- Dimension styles
- Text Styles
- Layers
- Blocks.

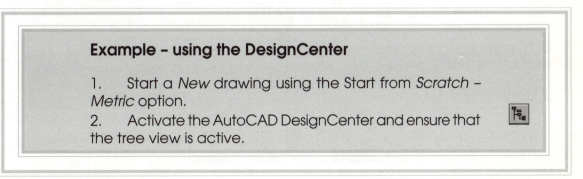

Example – using the DesignCenter

1. Start a *New* drawing using the Start from *Scratch – Metric* option.
2. Activate the AutoCAD DesignCenter and ensure that the tree view is active.

Inserting Dimension Styles

3.   Press the *Load* tool and select the drawing file 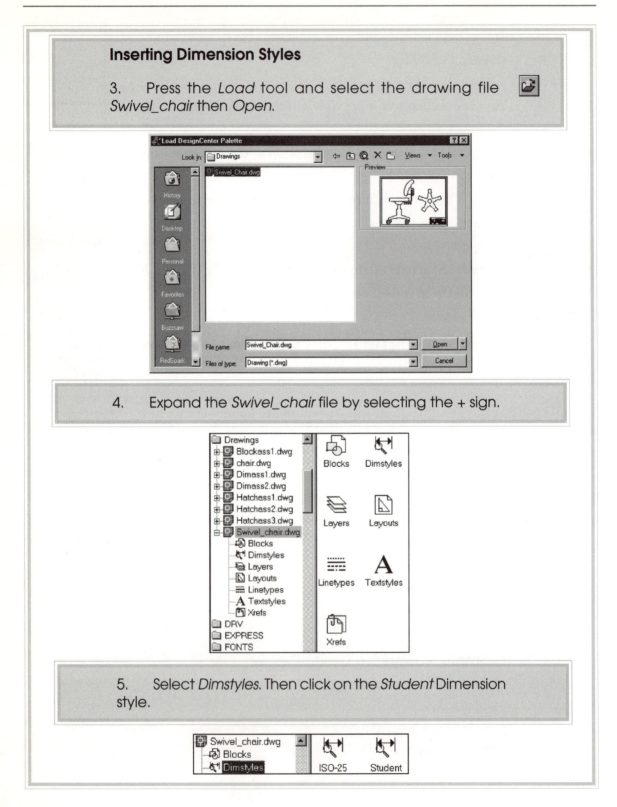 *Swivel_chair* then *Open*.

4.   Expand the *Swivel_chair* file by selecting the + sign.

5.   Select *Dimstyles*. Then click on the *Student* Dimension style.

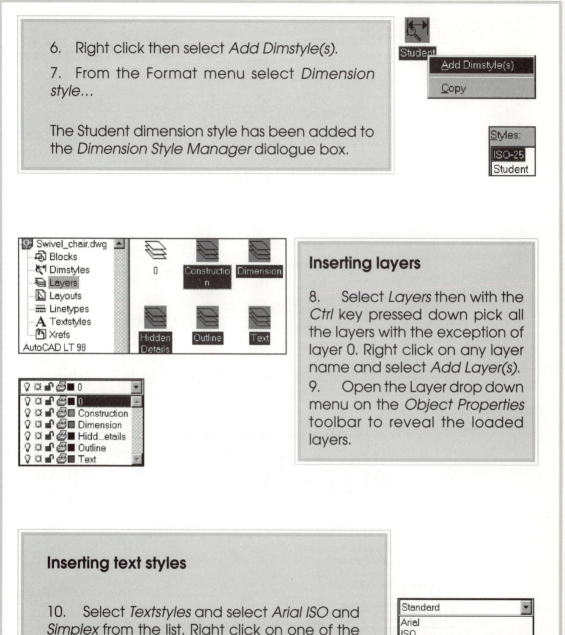

6. Right click then select *Add Dimstyle(s)*.

7. From the Format menu select *Dimension style...*

The Student dimension style has been added to the *Dimension Style Manager* dialogue box.

Inserting layers

8. Select *Layers* then with the *Ctrl* key pressed down pick all the layers with the exception of layer 0. Right click on any layer name and select *Add Layer(s)*.

9. Open the Layer drop down menu on the *Object Properties* toolbar to reveal the loaded layers.

Inserting text styles

10. Select *Textstyles* and select *Arial ISO* and *Simplex* from the list. Right click on one of the selected images and pick *Add Text Style(s)*.

To check that the text styles have been inserted, select *Text Style...* from the *Format* menu and reveal the style Name list.

Inserting blocks

11.   Press the *Find* tool.

Enter *Fasteners – Metric* in the *Search for the word(s)* edit box.

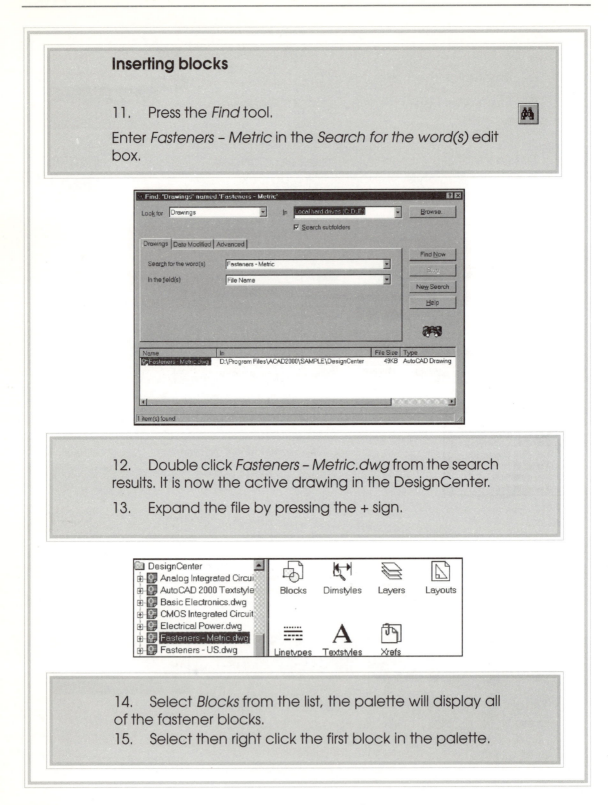

12.   Double click *Fasteners – Metric.dwg* from the search results. It is now the active drawing in the DesignCenter.

13.   Expand the file by pressing the + sign.

14.   Select *Blocks* from the list, the palette will display all of the fastener blocks.

15.   Select then right click the first block in the palette.

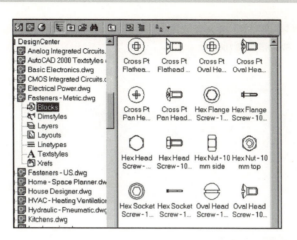

Note – the Insert Block... option will activate the Insert dialogue box and results in the block being inserted into the drawing.

Note – pressing the Cancel key will terminate the insertion of the block into the drawing but retain the block in the drawing database for insertion when required.

Note – the Copy option will add the block to the clipboard making it available for insertion into any drawing currently open.

16.   Select the *Insert Block...* option for the first ten blocks.

17.   From the *Insert* menu select *Block...* Press the down arrow to the right of the *Name:* box to reveal the list of available blocks.

Note – all the essential components are now available for you to create a new drawing or you may wish to save this drawing as a new template file.

18.   Save your drawing as *ADC.dwg*.

Chapter 19

Communicating between drawings

AutoCAD 2000i allows you to open several drawings simultaneously. This enables you to compare features within drawings and copy objects between drawings without having to close and re-open files.

You can also apply Hyperlinks which attach drawing objects to external files such as other drawing files, text documents, spreadsheets or even animations.

Objectives:

At the end of this section you will be able to:

▷   Open multiple drawing files.

▷   Copy objects between drawings.

▷   Create Hyperlinks.

▷   Complete a hyperlink exercise.

New commands

▷   Hyperlink

Opening multiple drawing files

You can have several drawings open at the same time in AutoCAD 2000.

1. From the *File* menu select *Open…* to activate the *Select File* dialogue box.
2. Navigate to the *Sample* folder within the *Acad2000 folder*.
3. Use the cursor to select *1st floor architectural*. Hold down the *Ctrl* key on the keyboard and select *1st floor electrical, 1st floor lighting* and *1st floor plan*.

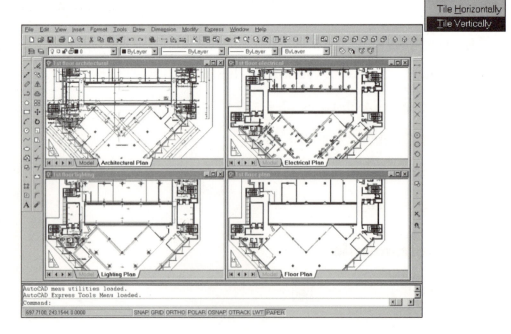

Each of the drawing files will highlight.

4. Press the *Open* button.

Each of the drawing files will load in turn and the last drawing selected will appear on the screen.

5. To view all the files simultaneously select *Tile Vertically* from the *Window* menu.

Example – copying objects between windows

1.   Notice that in each window, the layout tab is active indicating that the paper space environment is current. In each window click the *Model* tab to return to model space.

2.   Click into each window in turn and zoom a window around the central rectangular object.

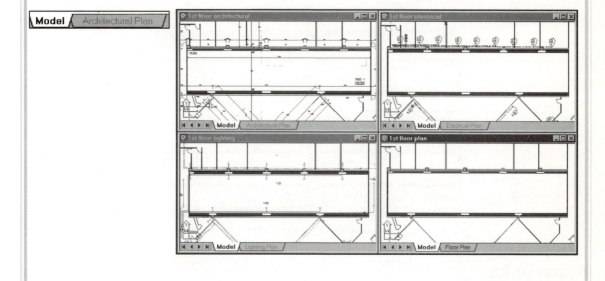

3.   In the top left window draw a rectangle and circle as indicated. Actual dimensional accuracy is not required. Position the objects approximately as shown.

4.   Select both objects to highlight them then right click to activate the cursor menu. Select *Copy with Base Point*.

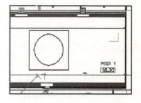

5.   Specify the base point as the centre (Osnap CEN) of the circle. Press *Esc* to clear the selection.

6.   Click into the upper right window to activate it then right click.

7.   Select *Paste to Original Co-ordinates*. The rectangle and circle will appear at the same co-ordinate position within the drawing.

8.   Repeat for both lower windows.

9.   Create other simple objects and use the cursor *Copy/Paste* commands to copy the objects to each of the drawings.

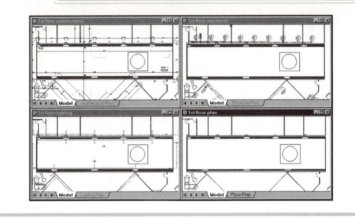

Hyperlinks

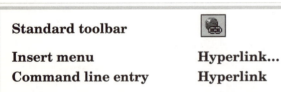

Standard toolbar	
Insert menu	**Hyperlink...**
Command line entry	**Hyperlink**

The Hyperlink feature allows you to attach a link to an object in the drawing which when selected will load an external file.

Examples of files that can be attached include another drawing file, text documents, spreadsheets or animations. To illustrate Hyperlinks you will create the following drawing that represents a simplified side view of a building. You will attach a hyperlink to the 1st floor annotation which when selected will load the drawing files specified.

Example - creating a hyperlink

1. Start a *New* drawing using the *Student - Template* option.

2. Create the following drawing. Do not include the dimensions and Array instructions.

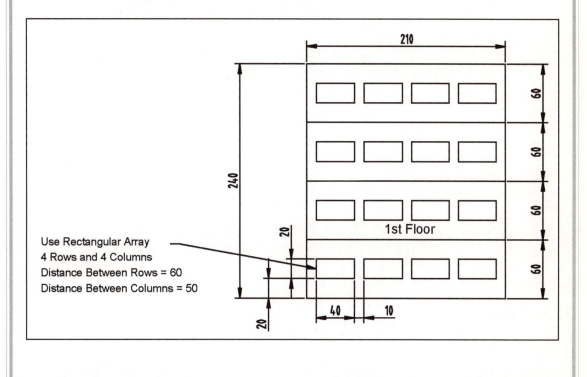

Use Rectangular Array
4 Rows and 4 Columns
Distance Between Rows = 60
Distance Between Columns = 50

210

240

60

60

60

60

20

20

40

10

1st Floor

Note – your drawing needs to have been saved before you can attach a Hyperlink.

1. Save your drawing as *Link1.dwg*.
2. Press the *Hyperlink* tool and select the text *1st Floor* to activate the *Insert Hyperlink* dialogue box.

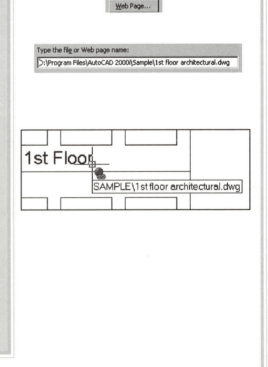

Insert Hyperlink

Link to:

Text to display:

Type the file or Web page name:
http://www.freeserve.com/

Existing File or Web Page

Or select from list:

Recent Files

F:\Butterworth\blocks2.dwg
I:\Drawings\Swivel_Chair.dwg
D:\Program Files\AutoCAD 2000i\Template\student.d
I:\Butterworth_Heinman\dwg.dwg
F:\Butterworth\blocks2.dwg

View of This Drawing

Browsed Pages

Inserted Links

E-mail Address

Path: http://www.freeserve.com/

Use relative path for hyperlink

1 object selected.
0 objects with hyperlinks.

Browse for:
File...
Web Page...

Target...

OK   Cancel   Help

3.   Press the *Browse for file...* button and navigate to the *Acad2000/Sample* folder.

4.   Select the drawing file *1st Floor Architectural* then press the *Open* button. The file name will appear in the *Link to file* edit box. Press *OK* to exit.

5.   To activate a Hyperlink you need to pass the cursor over the object to which the hyperlink is attached.

6.   Locate the cursor over the text 1st Floor. The Hyperlink icon will appear. Hold the cursor in position and the linked files name and location will appear.

7.   Select the text object to highlight it then right click.

Browse for:
File...
Web Page...

Type the file or Web page name:
D:\Program Files\AutoCAD 2000i\Sample\1st floor architectural.dwg

1st Floor

SAMPLE\1st floor architectural.dwg

8. Select *Hyperlink* at the bottom of the cursor menu then *Open SAMPLE\ 1st floor architectural.dwg.*

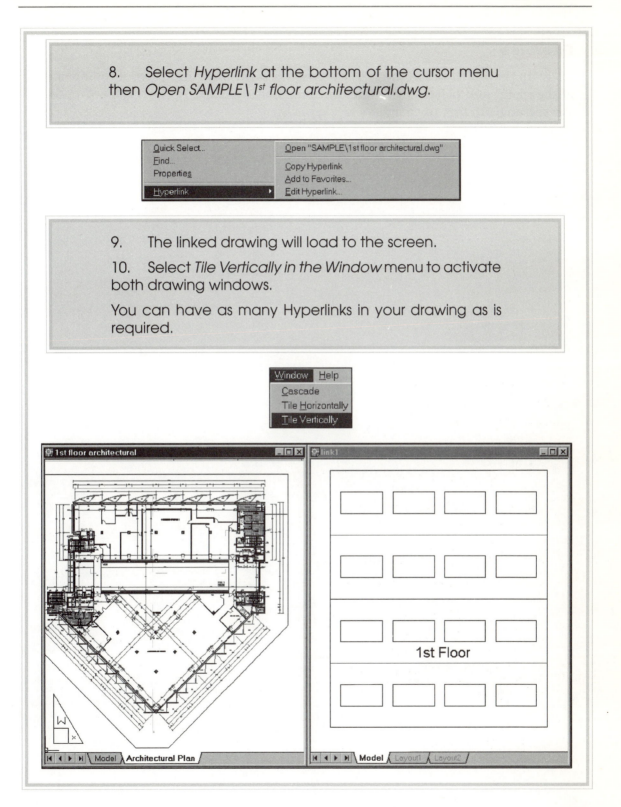

9. The linked drawing will load to the screen.

10. Select *Tile Vertically in the Window* menu to activate both drawing windows.

You can have as many Hyperlinks in your drawing as is required.

Exercise

Activate the *Model* tab in the *1ˢᵗ Floor Architectural* drawing. *Zoom* into any one of the rooms and place a rectangle that represents a chair. Apply a *Hyperlink* to the rectangle to link to your *Swivel_chair.dwg*.

Chapter 20

The User
Co-ordinate System

The User Co-ordinate System represents the current location and orientation of the X, Y and Z axes within the drawing. The X0, Y0, Z0 co-ordinate is by default, located at the lower left corner of the screen.

You can relocate this 0,0,0 co-ordinate to any location in the drawing at any time. This is particularly useful during the creation of large complex drawings.

Objectives

At the end of this section you will be able to:

▷   Relocate the current X0,Y0, Z0 co-ordinate to a new position.

▷   Complete a UCS exercise.

New commands

▷   User Co-ordinate System   UCS

The User Co-ordinate System

UCS toolbar	
Tools menu	UCS
Command line entry	UCS

The User Co-ordinate System (UCS) determines the position and orientation of the drawings X,Y and Z axis. Imagine your screen is like a sheet of graph paper. The X axis is along the horizontal edge and the Y axis is along the vertical edge. Objects are drawn and edited on this X,Y plane of the UCS.

The UCS icon can be displayed in one of two ways. Its default style is set to 3D. At the command prompt type UCS icon and press P for Properties.

This will activate the UCS icon dialogue box. Change the UCS icon style to 2D then press OK.

When you start a new drawing you will see the UCS icon at the lower left corner of the graphics area. The icon indicates the orientation of the X and Y axes. Where the W is present within the UCS icon this indicates that you are using the World Co-ordinate System, i.e. the default X,Y orientation and position when a new drawing is started. If the location or orientation of the UCS is changed then you are no longer using the original World Co-ordinate System and therefore the W will disappear.

The UCS plane can be located in any position or orientation in 3D space. This is crucial when creating 3D models. We are only concerned with 2D drafting at this stage therefore we will concentrate only on the UCS options that can be applied to 2D drawings.

Most drawing and editing commands are dependent on the position and orientation of the UCS. It provides the projection plane, for instance, when applying a relative or polar co-ordinate value when drawing a shape or moving an object, the orientation of the X,Y axis will determine the location of the next point specified.

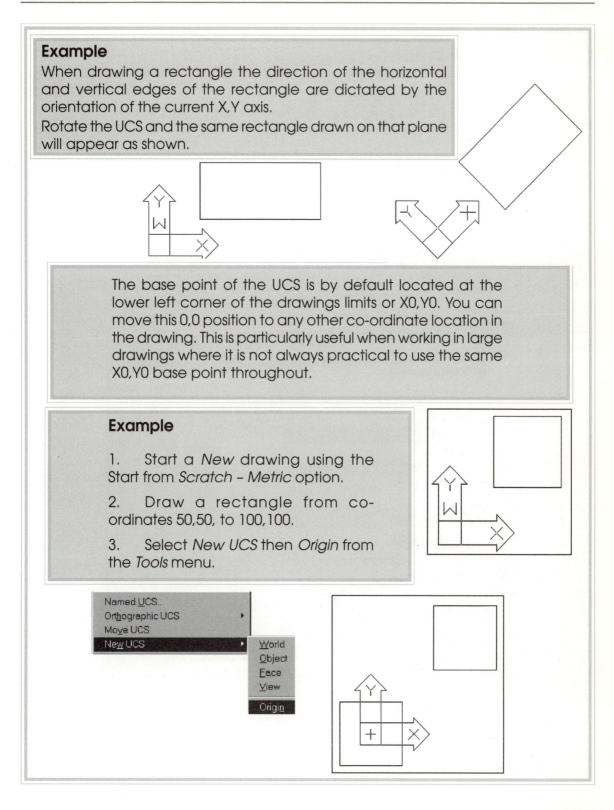

Example

When drawing a rectangle the direction of the horizontal and vertical edges of the rectangle are dictated by the orientation of the current X,Y axis.

Rotate the UCS and the same rectangle drawn on that plane will appear as shown.

The base point of the UCS is by default located at the lower left corner of the drawings limits or X0,Y0. You can move this 0,0 position to any other co-ordinate location in the drawing. This is particularly useful when working in large drawings where it is not always practical to use the same X0,Y0 base point throughout.

Example

1. Start a *New* drawing using the Start from *Scratch – Metric* option.

2. Draw a rectangle from co-ordinates 50,50, to 100,100.

3. Select *New UCS* then *Origin* from the *Tools* menu.

Named UCS...
Orthographic UCS
Move UCS
New UCS
World
Object
Face
View
Origin

4.   When prompted to specify a new origin point, enter 75,75.

The UCS icon will move to the new location and the W within the icon will disappear.

5.   Draw another rectangle from co-ordinates 50,50, to 100,100.

You will see that this co-ordinate is located in a new position.

You can create and name as many new UCSs as you need so that they can be restored at a later point.

6.   Select *Named UCS...* from the *Tools* menu. The current UCS has no name.

7.   Double click on *Unnamed* and enter *UCS1* then *OK*.

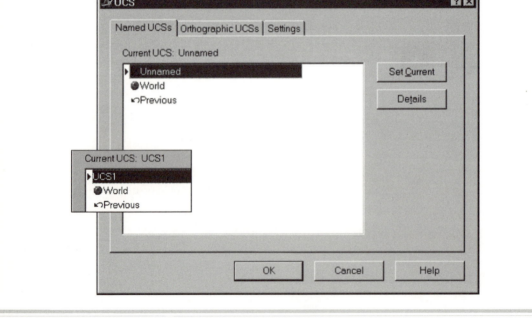

Chapter 21

Modelling using 3D solids

AutoCAD 2000i has a range of sophisticated 3D solid modelling and editing tools. Solid models are created by combining or subtracting solid shapes to and from each other. The solids toolbar provides six primitive solid shapes which may be used when creating a model. Alternatively, you can extrude or revolve a closed polyline to generate a solid shape.

This chapter will introduce you to the basic modelling and editing tools, which, through the completion of an exercise, will help you to understand more fully the potential of the solid modelling features.

Objectives

At the end of this section you will be able to:

▷   Understand the solid modelling and editing commands.
▷   Use the Extrude and Revolve features.
▷   Complete a solid modelling exercise.
▷   Relocate and reorientate the current UCS onto solid model faces.

New commands

▷	Box			▷	Extrude	EXT	
▷	Cone			▷	Revolve	REV	
▷	Sphere			▷	Union	UNI	
▷	Wedge	WE		▷	Subtract	SU	
▷	Cylinder			▷	Intersect	IN	
▷	Torus	TOR					

Modelling using 3D solids

Solids toolbar	
Draw menu	**Solids**

The Solids tools allow you to create solid models using primitive shapes. These primitive shapes can be joined together or subtracted from each other to form more complex models.

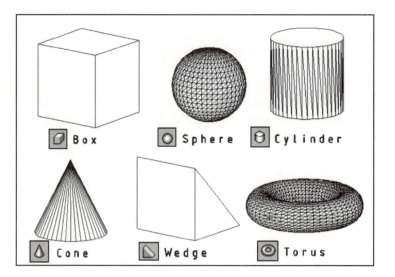

Objects can also be created by using the Extrude and Revolve commands to extrude (add thickness) or revolve closed polylines, rectangles, circles or ellipses.

Extrude command

Solids toolbar	
Draw menu	**Solids: extrude**
Command line entry	**Extrude**
Alias	**EXT**

The Extrude command is used to extrude to a selected object.

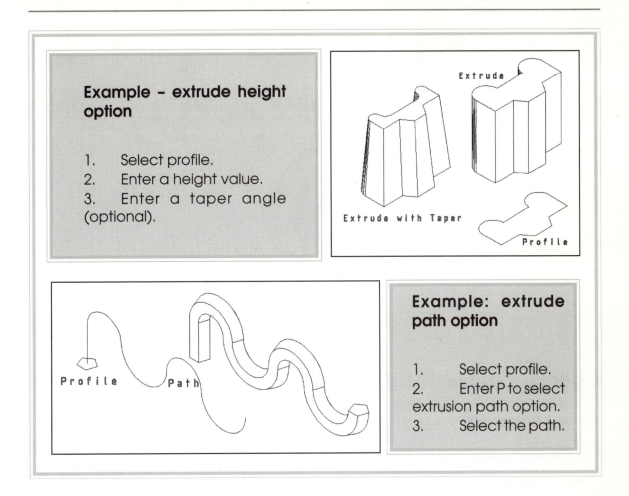

Example - extrude height option

1. Select profile.
2. Enter a height value.
3. Enter a taper angle (optional).

Extrude

Extrude with Taper

Profile

Profile   Path

Example: extrude path option

1. Select profile.
2. Enter P to select extrusion path option.
3. Select the path.

Revolve command

Solids toolbar	
Draw menu	Solids: revolve
Command line entry	Revolve
Alias	REV

The Revolve command is used to revolve a selected object about an axis

The axis of rotation can be a selected object or you can define an X or Y axis between 2 points.

The angle of revolution defaults to 360 degrees or you can specify a preferred angle.

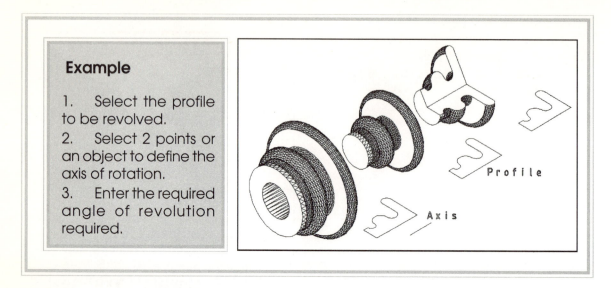

Example

1.   Select the profile to be revolved.
2.   Select 2 points or an object to define the axis of rotation.
3.   Enter the required angle of revolution required.

Modifying solids

Union command

Solids toolbar	⌀
Modify menu	Solids editing: Union
Command Line Entry	Union
Alias	UNI

The Union command will create a composite model from selected solids.

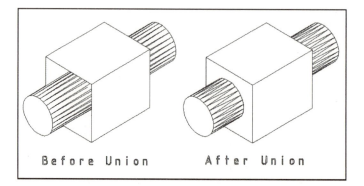

Subtract command

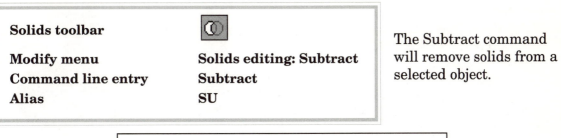

Solids toolbar

Modify menu   Solids editing: Subtract
Command line entry   Subtract
Alias   SU

The Subtract command
will remove solids from a
selected object.

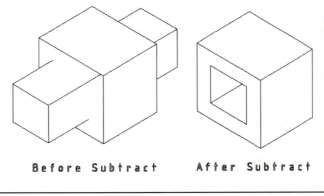

Before Subtract   After Subtract

Intersect command

Solids toolbar

Modify menu   Solids editing: Intersect
Command line entry   Intersect
Alias   IN

The Intersect command
will create a solid from
the overlapping volume of
selected solids.

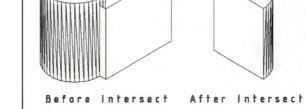

Before Intersect   After Intersect

Exercise using the solids commands

Instructions:

The object of the exercise is to create the following component.

Start a New drawing using the Use a Wizard, Quick Setup option set drawing units to millimetres and set your screen limits from 0,0 to 420,297

Note – remember to save your work at regular intervals.

1. Select the Cylinder tool from the Solids toolbar.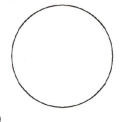

    Command: _cylinder

    Specify centre point for base of cylinder or [Elliptical] <0,0,0>: enter 100,100

    Specify radius for base of cylinder or [Diameter]: enter 50

    Specify height of cylinder or [Centre of other end]: enter 20

2. Select the Cone tool from the Solids toolbar.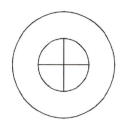

    Command: _cone

    Specify centre point for base of cone or [Elliptical] <0,0,0>: enter 100,100

    Specify radius for base of cone or [Diameter]: enter 25

    Specify height of cone or [Apex]: enter 100

3. Select the Cylinder button.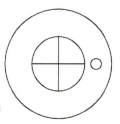

    Command: _cylinder

    Specify centre point for base of cylinder or [Elliptical] <0,0,0>: enter 135.5,100

    Specify radius for base of cylinder or [Diameter]: enter 5

    Specify height of cylinder or [Centre of other end]: enter 30

4. Select the Array tool from the Modify toolbar.

    Command: _array

    Select objects: select the small cylinder just created.

    Enter the type of array [Rectangular/Polar] <R>: enter P

    Specify center point of array: enter 100,100

    Enter the number of items in the array: enter 8

    Specify the angle to fill (+=ccw, -=cw) <360>: select default

    Rotate arrayed objects? [Yes/No] <Y>: select default

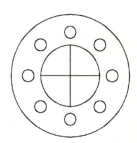

5. Select the Sphere tool.

    Command: _sphere

    Specify center of sphere <0,0,0>: enter 100,100,100

    Specify radius of sphere or [Diameter]: enter 25

6. Select the Torus tool.

    Command: _torus

    Specify center of torus <0,0,0>: enter 100,100,100

    Specify radius of torus or [Diameter]: enter 25

    Specify radius of tube or [Diameter]: enter 10

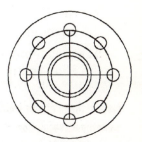

7. Create an isometric view of the component.

    Right click any tool button to present the AutoCAD toolbar list.

    Select View to activate the View toolbar. Press the SW Isometric View button

    At the command line enter HI to activate the Hide command to remove hidden lines.

    The next step is to use the modify tools to create the final component shape.

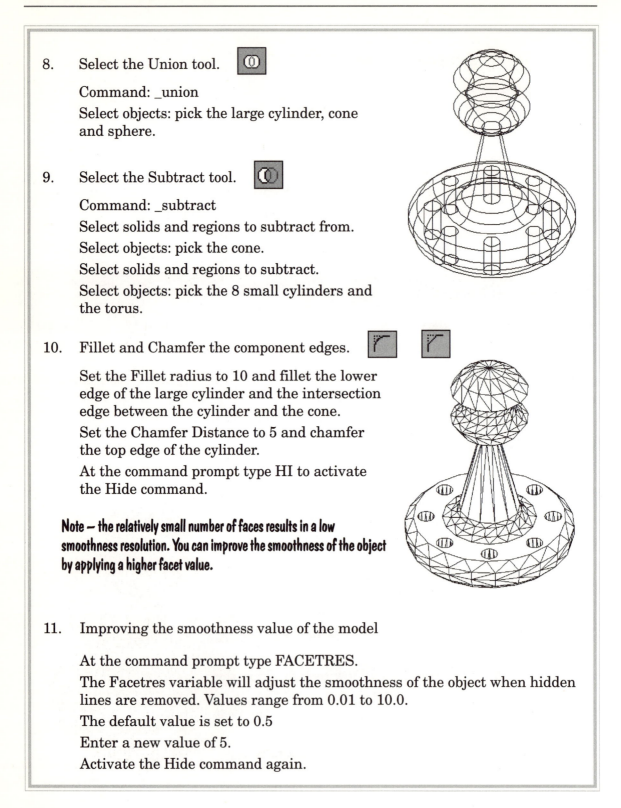

8.   Select the Union tool.

Command: _union
Select objects: pick the large cylinder, cone
and sphere.

9.   Select the Subtract tool.

Command: _subtract
Select solids and regions to subtract from.
Select objects: pick the cone.
Select solids and regions to subtract.
Select objects: pick the 8 small cylinders and
the torus.

10.   Fillet and Chamfer the component edges.

Set the Fillet radius to 10 and fillet the lower
edge of the large cylinder and the intersection
edge between the cylinder and the cone.

Set the Chamfer Distance to 5 and chamfer
the top edge of the cylinder.

At the command prompt type HI to activate
the Hide command.

**Note – the relatively small number of faces results in a low
smoothness resolution. You can improve the smoothness of the object
by applying a higher facet value.**

11.   Improving the smoothness value of the model

At the command prompt type FACETRES.

The Facetres variable will adjust the smoothness of the object when hidden
lines are removed. Values range from 0.01 to 10.0.

The default value is set to 0.5

Enter a new value of 5.

Activate the Hide command again.

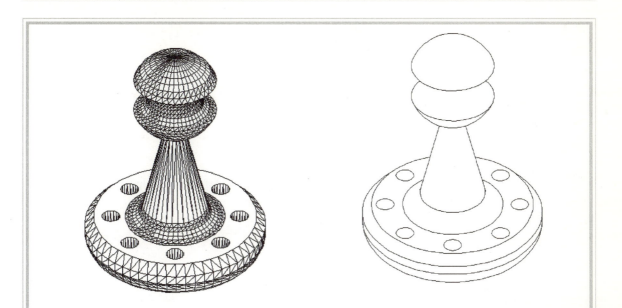

12. Further improve the smoothness of the component.

     You can remove the facets from your model and replace them with a smooth
     surface using the DISPSILH variable.

     At the command prompt type DISPSILH and enter a value of 1.

     Activate the Hide command.

13. Shade the component.

     Select the component to activate its
     grips. In the Object Properties toolbar,
     set the colour control to Blue.

     Right click any tool to present the
     AutoCAD list of menus. Select the Shade
     toolbar.

     Select the Gouraud Shaded tool.

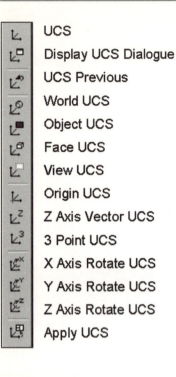

UCS
Display UCS Dialogue
UCS Previous
World UCS
Object UCS
Face UCS
View UCS
Origin UCS
Z Axis Vector UCS
3 Point UCS
X Axis Rotate UCS
Y Axis Rotate UCS
Z Axis Rotate UCS
Apply UCS

3D Modelling and the User Co-ordinate System

So far all your 2D drawings and 3D models have been created in the World UCS (see Chapter 20). In order to create more complex 3D models, the user will have to manipulate the World UCS so that it lies in any required plane or position.

The UCS feature has a variety of options that enable you to reposition the current X,Y plane.

Right click on any tool to present the AutoCAD toolbars list. Select UCS.

Orthographic UCSs

You can change the current UCS to a preset orthographic view using the Orthographic UCSs tab in the UCS dialogue box.

Press the Display UCS dialogue button.

Select the Orthographic UCSs tab.

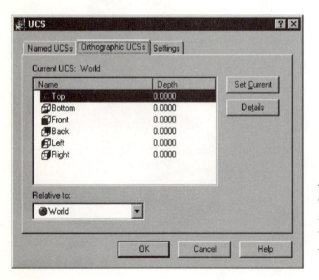

Alternatively, you can rotate the current UCS about an X, Y or Z axis to a specified angle, select an object's face to set the UCS or even use the current view as the UCS.

Exercise

We can best demonstrate this principle by completing the following 3D model exercise.

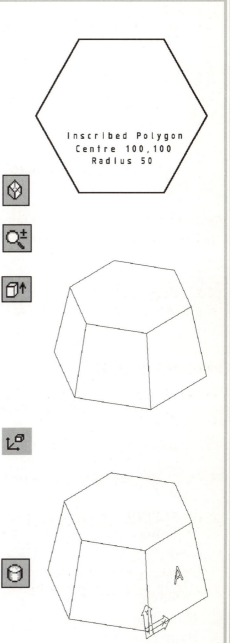

1. Start a New drawing using the Start from Scratch – Metric option.

2. Create the 6 sided polygon illustrated.

3. Select the SW Isometric View tool to view the polygon in Isometric.

4. Use the Zoom Realtime tool to reduce the size of the polygon on the screen.

5. Extrude the polygon to a height of 50 and a taper angle of 10 degrees.

6. Enter HI at the command line to remove hidden lines.

Your next task is to make the face A indicated the current UCS.

7. Select the Face UCS button in the UCS toolbar

8. When prompted to select the object face, pick a point close to the lower left corner of face A. The UCS icon will position at that corner and the UCS will align to face A.

9. Place a cylinder on face A.

command: _cylinder

Specify center point for base of cylinder or [Elliptical] <0,0,0>: enter 25,25

Specify radius for base of cylinder or [Diameter]: enter 10

Specify height of cylinder or [Center of other end]: enter -25

10. Subtract the cylinder from the extruded polygon.

Set The FACETRES variable to a value of 5. Type HI at the command prompt to remove hidden lines.

11. Select the Face UCS button in the UCS toolbar.

Pick the top face of the component at the point indicated in the illustration.

12. Select the Top View tool from the View toolbar to create a plan view of the object.

13. Draw a polyline profile using the co-ordinates below.

Specify start point: enter 20,15
@15<0
@15<90
@5<0
A (Arc)
@10<90
L (line)
@10<180
@5<90
@5<0
A (Arc)
@15<90
L (Line)
@15<180
Close

14. Select the SW Isometric View tool to view the polygon in Isometric.

15. Revolve the profile 360 degrees using the left side edge as the axis of revolution.

16. Subtract the revolved object from the extruded polygon.

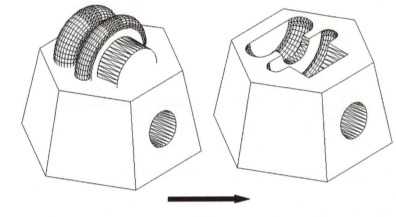

17. Select the component to activate its grips.

In the Object Properties toolbar set the colour control to Blue.

18. Right click any tool to present the AutoCAD list of menus. Select the Shade toolbar.

Select the Gouraud Shaded tool.

Practical assignment 6

Create the drawing illustrated. Name this drawing *Test1.dwg*. Do not include the dimensions or informational text. When you have created the drawing you will use the various modify commands to alter the appearance of the objects shown.

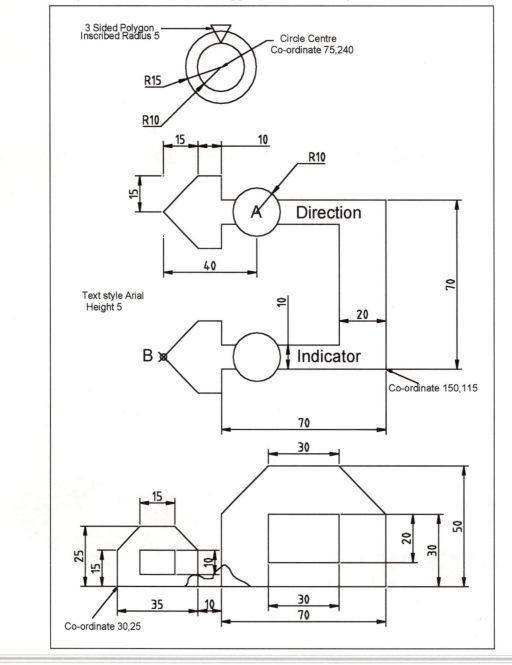

Drawing modifications

Drawing 1

1. Polar array eight copies of the 3 sided polygon around the circle.

2. Copy both circles and polygons a distance 55mm direction 0 degrees.

3. Trim the copied objects as indicated.

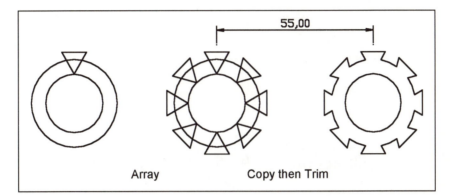

55,00

Array   Copy then Trim

Drawing 2

4. Mirror the drawing and text. Apply the appropriate variable to ensure that the text is not mirrored.

5. Determine the distance between the centre point A and the intersection point B.

6. Create a rectangle and insert the text Distance $A - B =$ and enter the distance value.

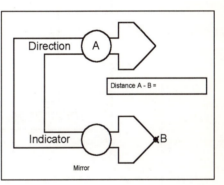

Direction   A

Distance A - B =

Indicator   B

Mirror

Drawing 3

7. Trim the lines and sketch.

8. Offset the house outline internally by 3mm.

9. Offset the rectangles internally by 2mm.

10. Save your drawing.

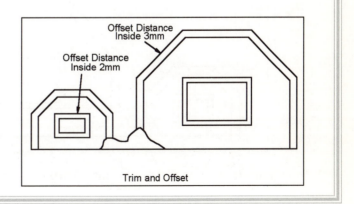

Offset Distance
Inside 3mm

Offset Distance
Inside 2mm

Trim and Offset

Completed drawing Test1.

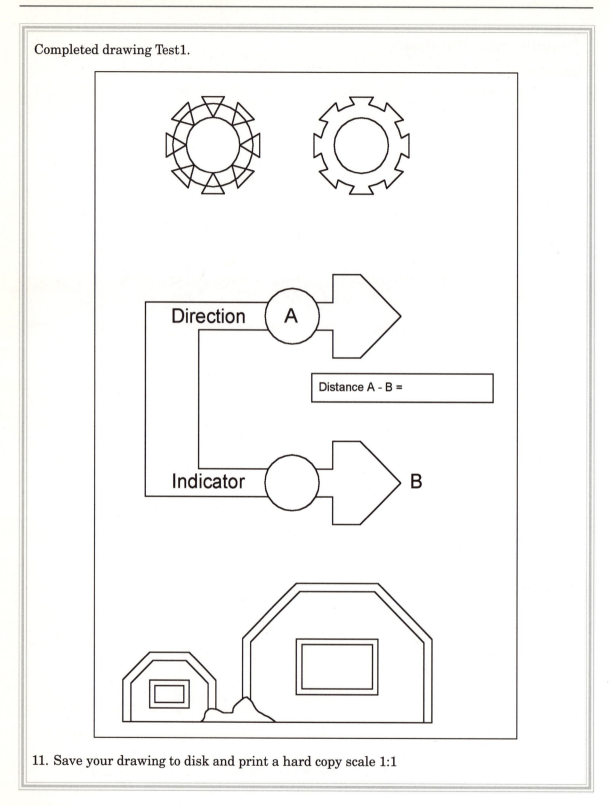

11. Save your drawing to disk and print a hard copy scale 1:1

Practical assignment 7

Create the drawing illustrated. Name this drawing *Test2.dwg*. Do not include the dimensions or informational text. Instructions will follow concerning the development of the drawing.

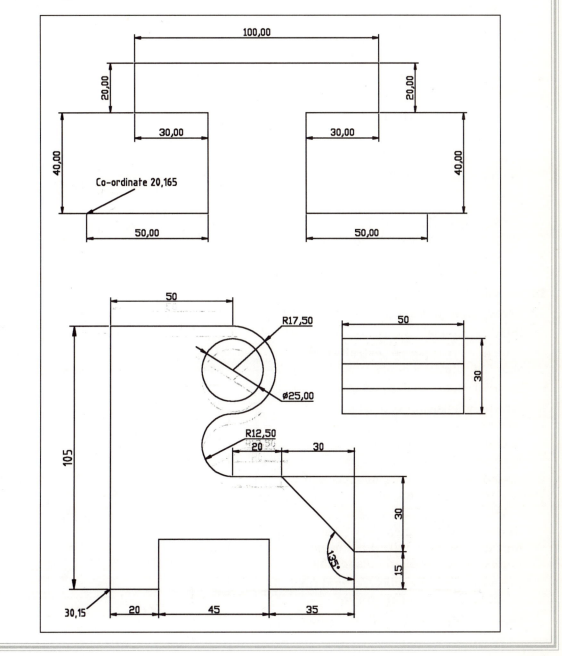

Instructions

Title block

1. Create a new text style called ISO and set to the following specification:
   Font: ISOCP.shx
   Height 0
   Width Factor 0.8
   Oblique Angle 0

2. Make the font style *ISO current*.

3. Set text height to 3 and enter *Title: Name:* and *Date:* at the left side of the Title box as indicated.

4. Set text height to 5 and enter your name and date *middle* justified as indicated.

5. Create a new text style called *Arial* and set to the following specification:
   Font: Arial
   Height 0
   Width Factor 1
   Oblique Angle 0

6. Make the font style *Arial* current.

7. Set text height to 5, *middle* justify and enter *Test2*.

Title:	Test 2
Name	A. Student
Date:	Feb 01

Dimensioned drawing

1. Create a new dimension style called *Test2*. Use the following specifications:
   ISOCP.shx font.
   Dimension text placed centrally and above the dimension line.
   Dimensions to two decimal places.
   Linear dimension text to follow dimension line.
   Radial and Diameter text forced horizontal.

2. Create a second dimension style called *Test2limits* using the following specifications:
   Dimension style copied from *Test2*.
   Linear dimensions toleranced using Limits. Set Limit tolerances to 0.5mm.

3. Create a new layer called *Testdims* and set colour to *Red* and make current.

4. Dimension the drawing as illustrated.

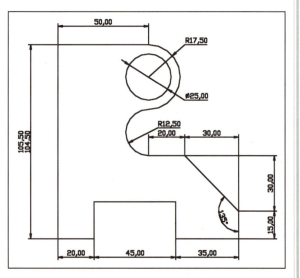

Road and houses

1. Move the UCS *Origin* to co-ordinate 20,165.

2. Create a new layer called *Road*, set its colour to *Blue* and make current.

3. Move the road outline to layer *Road*.

4. You are required to create the house block shown containing two attributes.

5. Draw the house outline illustrated. Do not include the dimensions.

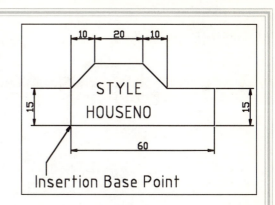

6. The specifications for the attributes are:

Attribute name	Mode	Prompt	Default value
Style	Constant-visible		Hilton
Houseno	Variable-visible	Enter House No:	

7. Set text height to 5 using the ISO text style.

8. Create a Block containing the house outline and both attributes. Name the block *House*. Set the insertion point at the lower left corner as indicated.

9. Complete the housing estate by inserting the blocks to the locations and angle indicated.

House number 1:	Co-ordinates 5,170	Rotation angle 0
House number 2:	Co-ordinates 15,260	Rotation angle 270
House number 3:	Co-ordinates 120,170	Rotation angle 0
House number 4:	Co-ordinates 175,200	Rotation angle 90
House number 5:	Co-ordinates 65,235	Rotation angle 0

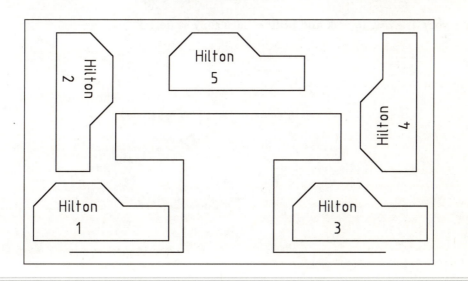

Completed drawing *Test2*.

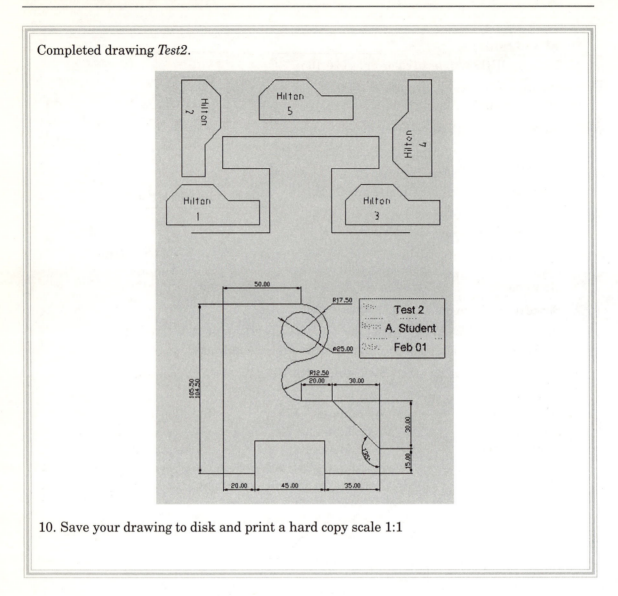

10. Save your drawing to disk and print a hard copy scale 1:1

Practical assignment 8

Template file

In this assignment you are required to create a number of objects placed at specified locations within a border. You will construct the drawings in model space but lay them out in paper space prior to saving and printing.

1. Start a *New* drawing using the *Start from Scratch-Metric* option. Set limits from 0,0 to 420,297.

2. Create the following *Layers* as specified below:

Layer Name	Linetype	Colour	Linewidth
Outline	Continuous	White	0.25
Title	Continuous	Blue	default
Construction	Continuous	Green	default
Psvports	Continuous	Magenta	default

3. Set the layer *Psvports* as the current layer.

4. Create the following text styles:

Style Name	Font	Width factor
Title	Arial	1.0
Headings	Simplex	0.8

5. Use the *Layout wizard* in the *Tools* menu to create a layout to the following specification:

Layout Name:	Test3
Paper Size:	A4 210x297mm
Orientation:	Landscape
Title Block:	None
Define Viewports:	Single viewport, Scaled to Fit
Location:	From 10,10 to 175,190

6. Make layer *Title* current and draw the border detailed below. Do not include the dimensions.

7. Create and locate the text within the title box. Include Title, Dwg No., Name, Scale and Date. Use text style Headings with a text height of 3mm.

8. Save the drawing as template file called *Test3.Dwt*.

275,195

Line Width 1mm

5,5

100

50.00

Title:

Dwg No: Name:

Scale: Date:

30

10

Creating drawings

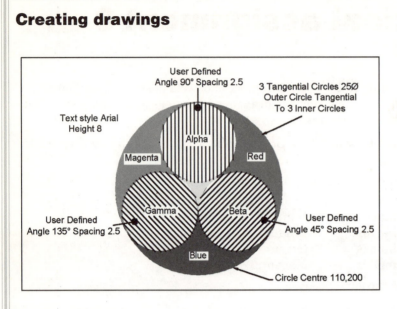

Text style Arial Height 8

Magenta

User Defined Angle 90° Spacing 2.5

Alpha

3 Tangential Circles 25Ø Outer Circle Tangential To 3 Inner Circles

Red

User Defined Angle 135° Spacing 2.5

Gamma

Beta

User Defined Angle 45° Spacing 2.5

Blue

Circle Centre 110,200

Hatched circles

1. Create the 25 radius circle.

2. Copy the circle 50mm direction 0 degrees.

3. Create a third circle using the *TTR option* tangentially touching the first two circles radius 25mm.

4. Create a final circle using the *Tan, Tan, Tan* option tangentially touching all three inner circles.

5. Move all four circles using the centre of the large circle as the base point, to co-ordinate 110,200.

6. Using the *Title* text style (Arial font) set height to 8mm and enter the text *Alpha, Beta* and *Gamma* centrally within the three inner circles.

7. Place the text *Red, Blue* and *Magenta*, height 8mm, centrally within the spaces between the 3 inner circles and the outer circle. Hatch these areas with the appropriate colour using the solid fill pattern.

8. Use the *User Defined* hatching option to hatch the three circles to the angle and spacing values indicated in the drawing.

Isometric object

1. Set *Snap and Grid* values to 10.

2. Activate the *Isometric Snap* style.

3. Draw the object indicated in the drawing. Do not include the dimensions.

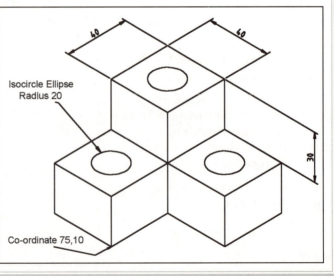

Isocircle Ellipse Radius 20

40

40

30

Co-ordinate 75,10

Gasket

1. Using the various circle options create the Gasket drawing illustrated. Do not include the dimension.

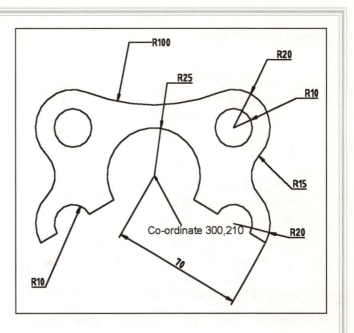

Test logo

1. Use the appropriate drawing and editing commands to draw the objects illustrated to the co-ordinates and dimensions shown.

2. The Polyline below the text is 1mm wide.

3. Use the text style Title (Arial font) text height 10, to create the text.

4. Use the appropriate text alignment option to make the text length 80mm.

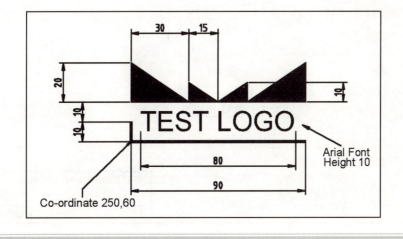

Paper space layout

1. Select the *Test3* layout tab to activate Paper space.

2. The objects will appear within the border as illustrated.

3. Make layer *Psvports* current.

4. Create a second viewport from co-ordinates 175,35 to 275,190.

5. Activate Model space within this second viewport and use the *Zoom and Pan* to adjust the viewing area so that the gasket fits the viewport.

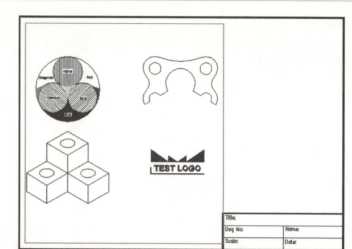

6. Return to Paper space. Freeze layer *Psvports*.

7. Enter the appropriate details into the *Title* box.

8. Save your drawing.

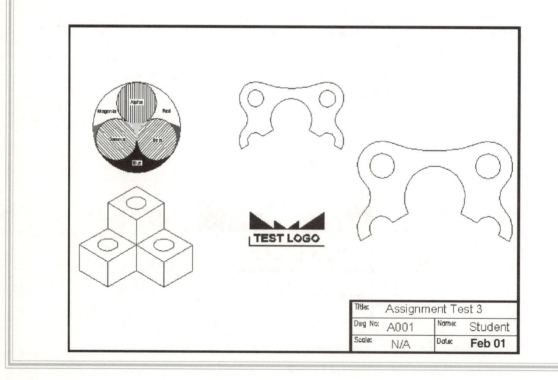

Multiple choice questions

1   The Object Properties toolbar cannot be used to:

A   Change an object's lineweight

B   Create a new layer

C   Change an object's colour

D   Load a new linetype

2   Which of the following options is the three Hatch command island detection styles?

A   Normal, Ignore, Centre

B   Outer, Pick, Select

C   Normal, Ignore, Outer

D   Normal, Centre, Outer

3   Which of the following commands would be the most efficient for converting drawing A to drawing B?

A   Copy, Scale and Move

B   Polyline with Osnap Parallel mode

C   Offset

D   Scale

4   Which of the following commands would be the most efficient for converting drawing A to drawing B?

A   Mirror

B   Extend

C   Move

D   Stretch

5   Which of the following options is the best way to produce a series of drawings each sharing common features such as layers, text styles, dimension styles, blocks etc.?

A   Use the DesignCenter

B   Create blocks and attributes

C   Create a template file

D   Start each drawing from scratch

6   An Associative dimension will:

A   Only apply to blocks

B   Modify its value and position when the object to which it is attached is modified

C   Automatically appear when the object is created

D   Dimension several objects at once

7   Which of the following best describes Attributes?

A   X,Y co-ordinate positions of objects

B   Text inserted using MTEXT command

C   Text applied to a dimension

D   Textual date within a block

8   Which variable is used to stop text being mirrored?

A   TEXTMIRR

B   MIRRTEXT

C   MIROFF

D   TXTOFF

9 Which of the following options will result in Lineweights appearing in the drawing?

A The current layer has a lineweight value

B Objects have a lineweight attached to them

C The LWT button is active on the status bar

D The variable LWT is set to 1

10 Which of the following options will reduce drawing size when an object is repeated many times?

A Multiple Array the object

B Multiple Copy the object

C Create one object then Mirror

D Insert the objects as Blocks

11 Which of the following commands would most efficiently produce drawing B from A?

A Break

B Trim

C Explode

D Erase

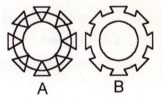

12 Which of the following options would most efficiently fillet the closed polyline object.

A Fillet – Select object option

B Multiple command then Fillet

C Fillet – Trim option

D Fillet – Polyline option

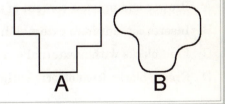

13 What is the purpose of a Layout?

A   A Layout is used to lay out multiple views of a drawing for plotting

B   To position circular objects within a polyline

C   Set drawing limits, layers and linetypes

D   Position multiple objects around the drawing in Model space

14 Which of the following statements is true?

A   The Ellipse – Isocircle option is only available if Isometric Snap is current

B   The Isocircle option is always available within the Ellipse command

C   Isocircles can be created when using rectangular Snap

D   Polar snap must be active to create Isocircles

15 The Hyperlink command allows you to:

A   Link objects together within the drawing

B   Link objects to external files

C   Link text to blocks

D   Link dimensions to objects

16 Which of the following statements concerning the AutoCAD DesignCenter is false?

A   Browse and preview content of external drawing files

B   Inserts content from external drawing files to the current drawing

C   Find blocks within external drawing

D   Create blocks from objects within the current drawing

Index

Absolute co-ordinate entry   22
Active Assistance   10
Aerial view   98
Arc command   109
Array command   78
Associative hatching   158
Associative dimensions   169
Attribute Definition dialogue box   220
AutoCAD Today   2
AutoCAD DesignCenter   242
Autodesk point A   8
Autosnap   53

Blocks   208
Block definition dialogue box   209
Block insertion   212
Boundary Hatch dialogue box   152
Break command   122
Bulletin Board   7

Chamfer command   75
Circle command   36
Command Aliases   15
Command line window   14
Construction line   102
Copy command   68

Direct distance co-ordinate entry   20
Dimensioning   162
Dimension styles   169
Dimension Style Manager   170
Dimension toolbar   162
Donut command   114

Drafting Settings dialogue box   101
Dynamic text   137

Ellipse command   111
Erase command   63
Extend command   65
Extrude command   264

Function keys   46
Fillet command   73

Grid mode   46
Grips   202

Hatching feature   152
Hatch Pattern Pallet   153
Help command   16
Hyperlinks   254

Insert command   114
Intersect command   267
Isocircle command   49
Isoplanes   48

Keyboard Shortcuts   59

Layer Properties Manager   124
Layer Colours   126
Layer Linetypes   127
Layouts   122
Lengthen command   121
Limits   101
Line command   20

Lineweight 128
Loading Linetypes 130
Linetype Manager 130
Linetype scale 131

Menus and menu bar 14
Mirror command 82
Model space and Paper space 228
Modify Object Properties 106
Modifying Multilines 107
Move command 72
Mtext 138
Multiline command 102

Object selection 62
Object snap 51
Offset command 70
Open Multiple Drawing Files 252
Osnap Settings tools 52
Ortho mode 46

Pan command 96
Polygon command 35
Polyline command 27
Polyline edit command 118
Printing and Plotting 236
Properties Toolbar 132
Point command 114
Polar co-ordinate entry 24
Polar mode 46
Polar Tracking 49
Properties command 196

Quick dimensioning 168
Quick Select 198

Ray command 102
Rectangle command 33
Relative co-ordinate entry 23
Revolve command 265
Rotate command 69
Running Object snap 52
Save Drawing As dialogue box 27

Scale command 120
Sketch command 111
Snap mode 46
Spell command 143
Spline command 110
Start a New drawing 9
    Using a wizard 10
Using a template 10
Start from scratch 9
Status bar 16
Stretch command 71
Subtract command 267

Templates 188
Text alignment 144
Text command 136
Text Control codes 139
Text editing 142
Text Style 136
Toolbars 15
    Customising 55
    New Tool Buttons 57
Tracking 24
Trim command 66

User Co-ordinate System 260
    Orthographic UCS 272
Undo/Redo commands 20
Union command 266
Units 189
Using Blocks and Attributes 118

Write block command 216

Zoom command 94